T0221178

KLIMAT

KLIMAT

RUSSIA IN THE AGE OF CLIMATE CHANGE

Thane Gustafson

Harvard University Press

Cambridge, Massachusetts ▪ London, England ▪ 2021

Library of Congress Cataloging-in-Publication Data

Names: Gustafson, Thane, author.
Title: Klimat : Russia in the age of climate change /
Thane Gustafson.
Description: Cambridge, Massachusetts : Harvard University
Press, 2021. | Includes bibliographical references and index.
Identifiers: LCCN 2021011244 | ISBN 9780674247437 (cloth)
Subjects: LCSH: Climatic changes—Russia (Federation) |
Climatic changes—Government policy—Russia (Federation) |
Fossil fuels—Russia (Federation) | Renewable energy sources—
Russia (Federation)
Classification: LCC QC903.2.R8 G87 2021 |
DDC 363.738/740947—dc23
LC record available at https://lccn.loc.gov/2021011244

This book is dedicated to Philip Vorobyov
Staunch friend, peerless colleague,
and keen-eyed observer of Russia

CONTENTS

A NOTE ON TRANSLITERATION

Transliteration from Slavic to Latin script is always a challenge, particularly where proper nouns are concerned. Throughout this book I have attempted to follow the guidelines of the US Library of Congress wherever possible. However, for place names the American Association of Geographers has its own conventions, and in most places I have followed those. (Thus: Ob River instead of Ob'.) Finally, if a place or person has been mentioned frequently in the Western media, then I follow the spelling used in the media. (Thus: Mikhelson instead of Mikhel'son.) Unfortunately, the result is a running series of compromises, which will make no one entirely satisfied.

KLIMAT

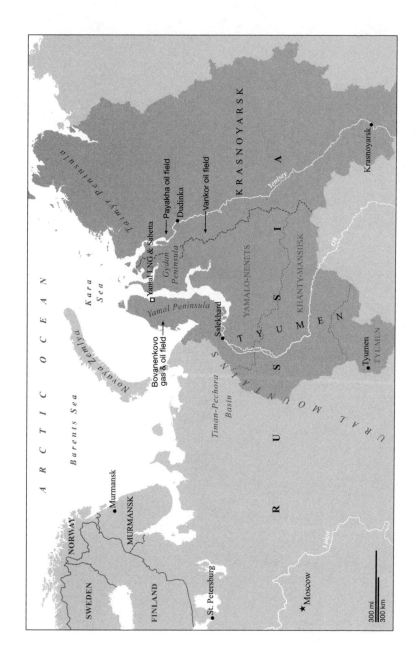

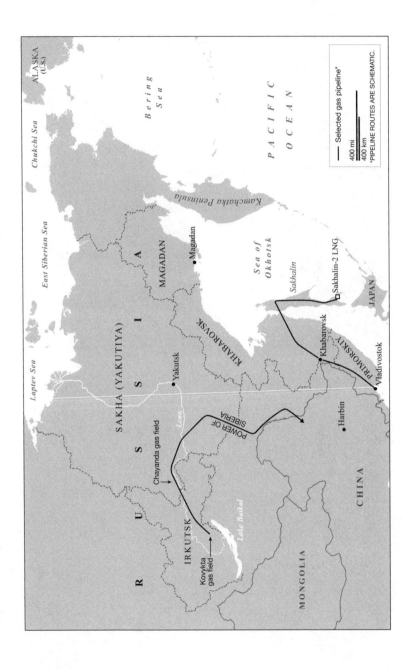

INTRODUCTION

Climate Change and Its Implications for Russia

Climate change is the defining issue of our century. It is at once the product of our civilization's greatest strengths and greatest weaknesses, of our unique ability to generate change and growth through innovation and enterprise, and of our inability to cope with the consequences. This is true for every country in the world. But for Russia the consequences will be especially dramatic, for its economy and environment, and for its standing as a great power in the rest of the world. That is the subject of this book.

As I write, the coronavirus pandemic has spread across the world, its consequences already far-reaching. As I see it, pandemics and climate change are two faces of the same problem. One is acute and the other chronic, but both are pathologies of an increasingly overcrowded and overburdened planet. COVID-19 is above all a disease of globalization, driven by the recent construction of world-spanning supply chains, catalyzed by the cheap and easy movement of goods, labor, and capital across national boundaries, accompanied by permissive state policies. So long as the uncontrolled contact engendered by globalization remains the dominant "business model" of the world, pandemics will remain a constant threat.

Yet even in the midst of a global viral outbreak, the chronic problem of climate change has not receded. As CO_2 concentrations continue to mount, climate change ticks away in the background, like an inexorable clock. Each year brings its message of growing damage—floods, fires, and heat waves—that briefly captures the world's attention, before being displaced by the next crisis—until the next reminder. As the decades go by, these symptoms will become more frequent and intense, with increasingly powerful political and economic consequences—crop failures and food shortages, deaths from heat waves, diseases of all kinds, and mass migrations.

Why is climate change so difficult to address? It challenges the very foundation of the economic growth model on which the stability of our political and social systems depends, and which has brought prosperity to many in just a few decades, while diminishing mass poverty around the world. No growth means no prosperity, and no prosperity means no stability. Climate change threatens all three, and in so doing, it undermines our most fundamental belief, shared by socialist and capitalist systems alike—our faith in unlimited human progress.

Not surprisingly, the effect is to divide us, not unite us. In a world in which there is no common power, we are unable to act together when our deepest interests conflict. As the consequences of climate change become more dire, the international politics of climate change will worsen toward even more disunity than today, for three reasons. First, climate change is the ultimate collective action problem. The benefits of limiting greenhouse gas emissions accrue globally, but the costs are borne by individual countries and communities. There will be little common ground among nations, especially when it comes to sharing costs. Russia, in particular, will face high costs and few benefits from joint action. The temptation to "free ride" will be irresistible.

Second, climate change sets the stage for confrontation, as extreme events engender new conflicts among and within countries. In many places, weak states will become completely unable to control their populations. The outcome will be growing anarchy and the threat of mass migrations, accompanied by cross-border violence and xenophobia. Even in Russia, where the state is likely to remain strong, social changes aggravated by climate change, especially migration, will pose mounting challenges.

Third, the combined result of these conflicts will be changes in the distribution of wealth and power. Some countries and groups will benefit from climate change, whereas most will be harmed; the poorest and most vulnerable will face acute shortages. There will be winners and losers, and these changes will likely not be peaceful. Russia, I argue in this book, will be one of the losers.

These features are already all too apparent today, especially in the continuing failure of most of the world's elites to move beyond declarations. Russia's politicians and business leaders are no exception. As I will show, although awareness of climate change has risen among both Russian elites and the public in the past few years, it has yet to be translated into meaningful policy.

These are the issues that have led to this book. But why is Russia so significant? For much of our lifetimes, one of the dominant narratives—at most times *the* dominant narrative—has been the rivalry between Russia and the West and the two opposing political and economic systems they represented. For a time, the advent of liberal democracy as the new world order—the "end of history"—was our hope and belief. But in Russia that vision was never shared by more than a small minority, whereas most Russians saw the 1990s as a disaster and a humiliation. In 2000 Vladimir Putin came to power with the self-appointed mission of restoring Russia's standing as a great power. Now, twenty years on, Russia and the West have reverted to what amounts to a new cold war, threatening positive achievements and raising new dangers.

But by mid-century the entire context of great-power rivalries, the rise and fall of political leaders and movements, the tidal succession of fashionable ideas, and above all the shared obsession with competitive growth and military power, will be increasingly overshadowed by the growing reality of climate change. Hence the questions of this book: How will Russia's territory—as well as its political system, its economy, and its society—be affected by climate change? How will these changes alter Russia's standing as a great power? What, indeed, will be the sources of "great powerdom" by 2050? Will Russia's future role in the world economy enable it to compete as a great power? And how will Russia react if it cannot?

What happens in Russia will matter very much to the rest of us. Russia is the fourth largest emitter of greenhouse gases in the world. It has the world's largest reserves of oil and gas, exceeding even those of Saudi Arabia. It has the most successful civilian nuclear power program currently in existence, a valuable potential resource in the event of a rebirth of nuclear power later in the century. Its agricultural renaissance promises a vital contribution if crops fail in the Middle East. Russia's actions on the international stage will help shape, whether positively or negatively, the course of the world's policies to deal with climate change. What Russia does will be very important.

But Russia's situation at this moment is paradoxical. Over the last twenty years Russia's economy has made considerable progress compared to its low point after the collapse of the Soviet Union in 1991. Its oil industry has been modernized through new technologies. Its gas industry has developed a new generation of supply, and it has built a new network of gas pipelines to transport it. A private-sector start-up is developing liquefied natural gas (LNG) to export to Asia. Its nuclear power sector has been rebuilt and has become a formidable force in world markets. Coal and metals have been revitalized. The agricultural sector has been privatized and is turning Russia into an export powerhouse. Its telecommunications system has been modernized. Russia's fiscal and monetary system has been rebuilt and its finances are strong. In general these changes have not been adequately recognized in the West.

Yet with few exceptions, the changes that have occurred mostly conform to a traditional Russian model—a preference for large companies that, even when they are private, are intertwined with the state in a web of interdependence. The state remains the dominant player, as it has throughout Russian history. The pattern of policy remains top-down rather than bottom-up. The net effect is to strengthen Russia's traditional industrial model, and especially its reliance on exports of raw materials, above all hydrocarbons. The paradox is that the very changes that have improved Russia's economy over the past two decades ultimately make it more vulnerable to—and less able to adapt to—the challenges of climate change.

The Challenges of Climate Change for Russia

Russia is near the top of all the countries that stand to be affected by climate change. It is one of the countries that depend most on the export of fossil fuels. One-third of its territory lies north of the Arctic Circle, and its Arctic coastline stretches for 24,000 kilometers, much of it consisting of permafrost, an unstable mixture of sand, ice, and methane. Its forests, vulnerable to disease, drought, and fire, are the world's largest, and account for over half of Russian territory and one-fifth of the world's forested area. And as noted, as the world's fourth-largest emitter of CO_2 and many other harmful pollutants, Russia is already one of the chief causes of climate change; but as time goes on, it will also be one of its chief victims.[1]

Climate change will alter the Russian landscape. Large areas of arctic permafrost will melt, endangering local infrastructure such as pipelines, making houses and buildings more fragile, and making new construction more difficult and expensive. The warming of East Siberia will bring increases in pest-borne diseases. Forest fires will become more frequent and extensive. Areas of marginal rainfall, covering much of south Russia, will experience more frequent droughts, and agricultural production, which has been a success story in recent years, will be affected as well.

Not all of the consequences of climate change will be bad for Russia. Higher average temperatures will cause Russian agricultural yields to increase in some places. Russia's northern coastline will become a transit route between Europe and Asia, as arctic ice melts and year-round passage enables exports of oil and LNG to the east. Outside Russia, there may be increased demand for nuclear power, and Russian technology will be well placed to provide it. Food shortages, especially in the Middle East, will open up new export markets, which will also bring geopolitical leverage. But on balance, I argue here, the net effects will be negative.

Much of this book is about energy. Energy, especially oil, is at the center of Russia's challenges from climate change. The combination of technological advances on many fronts, tighter climate regulation, and the changing attitudes of investors and the public, will likely bring a peak in global oil consumption, perhaps as early as the mid-2030s, followed by a steady decline, together with lower prices. This in turn will cause a fall in Russia's income from oil exports, which account for between one-quarter and one-third of the Russian federal government's revenues, depending on annual world prices. Income from natural gas and coal will not be sufficient to fill the gap. As a result, Russian economic growth will slow, and the ability of the state to finance its welfare system and state-led investment, and to act as the economy's provider of last resort, will weaken.

But energy from fossil fuels is only half of the story. The other fundamental question addressed in this book is whether Russia's revenues from non-fossil-fuel and non-energy exports will offset the decline in revenues from fossil fuels. Can exports of metals help to make up the difference? Can agricultural exports continue to grow? Can Russia become a major exporter of civilian nuclear technology, as it has the ambition to do? Can the development of new transportation routes turn it into a profitable transit hub between Europe and Asia? All of these questions are explored in this book as well.

To assess the overall impact of climate change on Russia, we must distinguish between direct and indirect effects, and between external and internal ones. Direct internal effects are those that affect the domestic economy and the population inside Russia—for example, the impact of global warming on crop yields, coastal infrastructure and transportation, and standards of living and public health. Indirect external effects, such as changes in peak oil demand, arise from changes in the world economy as a result of climate change, which in turn will affect Russia's position in international flows of trade and technology.

The key argument of this book is that the indirect external effects, and above all trends in the global energy transition, will be the chief determinants of Russia's future wealth and power by 2050. Then the direct internal effects, chiefly in the Arctic and in Russia's southern agricultural zone, will become increasingly dominant as the century moves on. But the main horizon of this book is 2050.

The Impact of Climate Change:
Russia vs. the United States

Compare Russia's situation with that of the United States. In some respects Russia is more fortunate. Both have long coastlines, but whereas the three main US coasts (including the Gulf of Mexico) are densely populated and are major commercial and manufacturing centers, the Russian coastline is largely unpopulated and undeveloped, with the exception of the region around Saint Petersburg and the Black Sea.[2] A significant portion of the Russian extractive sector lies in arctic areas, not far from the coastlines, but only about 6 percent of the population lives there. Murmansk and Vladivostok and Norilsk cannot compare in size or importance with Miami and New Orleans, let alone New York.

It happens that because they lie in the path of major storm systems, the US East Coast and Gulf Coast are highly vulnerable to climate change, more so than the corresponding coastal regions of Russia. In particular, the coast of Florida, the fastest-growing state in the United States, is increasingly exposed to flooding by rising ocean levels. Miami is rated as one of the most vulnerable cities in the world. New Orleans, because of increasing river flooding and storm surges, is equally at risk, and the New York / New Jersey region is not far behind.[3] In Russia, despite the growing dangers posed by melting permafrost along its long northern coastline, there is nothing comparable.

The two countries also face different risks from forest fires. In places such as California, as many as one-fifth of all homes are at some degree of risk. Even now, losses from forest fires threaten the insurance industry, which has responded by red-lining the most exposed areas and refusing to insure them. This in turn not only threatens many homeowners with major losses, but prevents them from obtaining mortgages or selling their homes. Russia faces nothing remotely like this, especially in the sparsely populated north and east, where forest fires burn over comparable areas but without the same economic consequences.[4] Forest fires in Siberia and the Far East, though they consume millions of hectares of remote taiga every year, do not threaten population centers or even exports of timber.

The same is true of the outlook for agricultural production in the two countries. The middle of the United States—the corn, soy, and wheat powerhouse of the world—is already suffering from a combination of flooding and drought. In the Midwest, spring floods in the Missouri and Mississippi basins are delaying spring planting and causing mounting crop losses. Farther west, where agriculture is based on irrigation, more frequent and severe droughts are causing farmers to drill their wells deeper, making the underlying aquifer sink faster. By 2050, large areas of American agriculture will have become unprofitable, despite the vast subsidies that go to the US farm sector. Again, there is nothing comparable in Russia.

In short, the United States already faces a mounting burden of difficulty and expense to adapt to the direct internal effects of climate change, while Russia does not, at least not on the same scale. Yet the United States will enjoy major advantages over Russia in its capacity to adapt. One example is the role of the financial sector, which is much more highly developed in the United States than in Russia. Financial markets focus on the leading edge of the market, where technology has its most disruptive effects. Financial players are quick to respond to opportunities at the margin, such as "shorting the losers" when coal prices go down, or buying carbon emissions certificates as investment vehicles. Once a downward trend is under way, financial players are quick to bail out, causing the capitalization of incumbents to collapse, leading in short order to delisting, massive restructuring, or even bankruptcy. These responses, by creating powerful market signals, accelerate the transition for the energy sector as a whole, compared to changes in regulation, which may take years to debate and put into effect.

Other features of the American economy also work to the advantage of the United States compared to Russia. Because overall investment is lower in Russia than in the United States, the existing Russian stock of capital and infrastructure will be replaced more slowly. The greater state role in energy regulation will prevent price signals from accurately reflecting fast-moving cost trends in areas such as renewables and power storage. State bureaucracies will be slower to respond than private players. Technological "leapfrogging" will be held back in a system in which there is less start-up entrepreneurship. For all these reasons, the energy transition will be slower in Russia than in the United States, in both domestic and export policy.[5]

Because more than half of the revenues of the Russian government come from traded natural resources, an important variable will be the degree to which the world economy remains open. If trade barriers go up, the cross-border transfer of new energy technologies will slow. A lower rate of economic growth worldwide—a predictable consequence of a more closed global economy—will diminish the rate at which countries, especially developing ones, will renew their infrastructure and industry and transition toward greater energy efficiency. The longer the global energy transition takes, the longer Russia's oil-export revenues will remain strong. But this will only prolong Russia's continued dependence on hydrocarbon exports and delay its adaptation to new roles in tomorrow's world economy.

Climate Change and COVID-19

How are these fundamental questions likely to be affected by COVID-19? Will the pandemic be a game changer? It is probably more correct to think of the virus as a catalyst that will accelerate larger trends that were already under way before the pandemic. The following three aspects seem especially significant:[6] First, as the world begins to emerge from COVID-19, the capacity and the will of governments for concerted action, already weak to begin with, will weaken further, as governments focus inward and concentrate on their own independent strategies for rebuilding their economies and their societies. Instead of the concert of nations, we are more likely to see suspicion and hostility, as countries blame one another for the crisis and redirect their lending and aid to their own populations.

Second, the world will be deep in debt, and will likely remain so for decades. This will curtail financial assistance to the worst-off. As governments, companies, and consumers struggle to repair their balance sheets, meet interest payments, and work off what they owe, there will be little capacity or appetite for anything but the most urgent tasks. Proposals such as financial aid from developed countries to developing ones to combat climate change will likely be off the table, displaced by more immediate tasks. This will be a world of little generosity.

Third, the global flow of capital, goods, and high-tech entrepreneurship, which never fully recovered from the financial crisis of 2008–2009, will remain weaker than it was on the eve of the pandemic. Tariff barriers and protectionism will be the norm, as governments seek to protect their industries and maintain employment. Cross-border investment will be slower. Global investments in "clean energy" could decline and ambitious targets for "carbon-neutral economies" be put off. Insofar as there is still a clean-energy agenda in Asia, it will be more directed against pollution—that is, the issue will be public health rather than reduction of greenhouse gas emissions.

These three features will apply to Russia as well. Russia was never as fully integrated into the globalized economy as the other major powers. It was only in the last decade that Russia became fully connected to the internet, and even then primarily in Moscow and Saint Petersburg. Its participation in global financial flows remained limited, even before the financial crisis of 2008–2009, and the net movement of capital was outward instead of inward. Russia was only a minor participant in global outsourcing—for example, in call- or data-processing centers—and it had no large surplus of cheap labor to draw upon in the international division of labor. Its comparative advantage in international trade resided primarily in the export of raw materials. In a "deglobalizing" world, these features would likely become even more pronounced.

Yet the future course of the pandemic and its longer-term consequences are so uncertain that one can hardly do more than speculate. For that reason, in the rest of this book I will mostly sidestep COVID-19. Pandemics may come and go, but climate change will remain a steadily growing force. That is our subject here.

Climate Change and the Energy Transition

At this book's center is the concept of "energy transition." After more than a century of research, debate, and increasingly rigorous modeling, the chief cause of climate change is by now abundantly clear. The consensus among climate scientists is all but unanimous that climate change is due to the greenhouse effect engendered by man-made emissions of CO_2 and methane, largely as the result of the combustion of fossil fuels—oil, gas, and coal.[7] Thus, in broad historical perspective, the message of climate change is unmistakable: it signals the approaching end of fossil fuels as the basis for our civilization. The end will not come in this generation, or in the next. We are barely at the beginning of the beginning. But the era of fossil fuels is closing.

There have been other energy transitions in human history, as coal displaced wood, and oil and gas then displaced coal, while all of those fuels displaced animal and human power. The two key lessons from the past are that energy transitions take many decades; and that the old and the new coexist for a long time, until the transition is finally complete.[8] So far, the latest transition, to electric vehicles and renewable sources of power, has gone much faster than preceding ones.[9] But fossil fuels will continue to dominate for several decades more, although the mix will change—oil and coal will go down, gas will go up, renewables will boom, in varying proportions. But by 2050, the share of fossil fuels will be declining worldwide.

"Energy transition" pushes in two opposite directions. Driving it are two powerful forces—technology, and the inexorable growth of human numbers and expectations. The effect of technology is two-sided: On the one hand, it has brought successive revolutions in oil and gas— first shale gas and then tight oil—which have lowered costs and boosted the production of hydrocarbons. On the other hand, technology also is driving a steady and dramatic decrease in the costs of renewables, so that new solar and wind in many places are now cheaper sources of electricity than existing power plants using fossil fuels.[10] Today the first effect is still dominant; but as the decades go by, the second will become steadily more powerful, because it leads to a steadily growing share of electricity in final energy consumption. The chief battleground in the energy transition will be electricity, as digitalization makes more things electric.[11]

The other major driver of the energy transition is the demand of the world's growing urban populations for rising living standards. The United Nations Human Development Index, a composite of several measures of popular well-being, shows a clear relationship between increasing well-being and energy consumption, up to about 100 gigajoules per year. At present, 80 percent of the world's population consumes less than that.[12] The demand for more energy, especially from the developing world, will be inexorable. The only question is how it will be satisfied.

Opposing these driving forces is the tremendous inertia of existing energy systems, of which the best symbol is the rearguard action being fought by coal. Coal is retreating worldwide, yet it is a slow process.[13] The best illustration is China, where coal is still the leading energy source. The Chinese coal industry is deeply entrenched in the economies of entire provinces, which depend on it for employment.[14] China today continues to build coal-fired power plants, but it is not alone. Across the rest of the world, according to a study by the London-based Overseas Development Institute, the governments of the G20 countries spend nearly $64 billion per year to support the production and development of coal, of which almost three-quarters goes to coal-fired power generation, including new construction—and the total is growing steadily. It will be a long time before coal is squeezed out of the world energy system, although some studies indicate that coal demand could peak as early as 2022.

The speed of the energy transition is thus crucial. But it is surrounded by many uncertainties, most of which, I will argue, lie outside of Russia's control. What Russia does, or does not do, will have little impact on the course of the energy transition in the rest of the world. What is more important is the impact of energy transition on Russia.

The consequences of the energy transition for Russia will unfold in two phases. During the first phase, roughly to the beginning of the 2030s, global energy demand, including for hydrocarbons, will continue to increase, and Russia's energy exports will remain strong. Oil prices may even increase from their present low points, and there will be occasional price spikes as in the past. Russia's gas and coal exports will grow. Russia will have a decade of respite, during which its basic industrial and political model, founded on energy exports, will remain temporarily viable.

In the early 2030s, however, a second phase will begin. As the energy transition takes hold worldwide, Russia's export revenues from oil, gas, and coal will decline sharply, with corresponding pressure on the economy, the society, and the state. Russia's revenues from other sources, such as agriculture, will not be able to compensate. This will mark a major turning point for Russia. From that point on, the external and internal impacts of climate change on Russia will only increase.

Russia's Voyage to 2050

This book's horizon is 2050. This may seem a long way off, yet it will be upon us before we know it, a mere blink of an eye. What will we see? On present trends, despite the plethora of political pledges, the world will have missed its targets for greenhouse emissions by a wide margin. Instead of "zero carbon," annual CO_2 emissions will have grown from today's 35 billion tons to over 50 billion.[15] The world will have warmed, not by the 1.5 or 2 degrees Celsius that it has risen since the beginning of the Industrial Revolution, but perhaps by 3 to 5 degrees, particularly at earth's northern latitudes, where most of Russia is located. In such a world, climate change, and the world's increasingly desperate responses to it, will have become the dominant issues in global affairs, surpassing all others. Yet the desperation of politicians and public opinion will not necessarily accelerate the energy transition. For Russia, in particular, the crucial question will be the world's demand for oil and gas; and that will depend, as always, on the complex interaction of trends in policy, technology, and economics, not on media events. Therefore I take a cautious view on the rates and dates at which hydrocarbon demand will peak, but the premise of this book is that the tipping point is coming, most likely in the 2030s.

This book consists of four parts. Chapter 1 recounts the rising awareness of climate change among Russian climate scientists and its gradual translation into a political and public issue. Chapters 2–4 address the energy transition and cover the three fossil fuels—oil, gas, and coal. Chapters 5 and 6 discuss Russia's energy alternatives to fossil fuels—renewables and nuclear power. Chapters 7–9 take up three other aspects of Russia's adaptation to climate change and possible alternative sources of export revenues—agriculture, the Arctic, and metals. The Conclusion sums up the challenges of climate change to Russia's fossil fuel model and looks ahead to the prospects for Russia's broader adaptation.

To sum up my argument in this book: First, the world of climate change will be highly unfavorable inside Russia, where temperatures are already rising two-and-a-half times faster than in the rest of the world. This rise in temperature is already having two main internal consequences—melting permafrost and more frequent extreme events (droughts, floods, and heat waves) affecting cropland. There will be only one positive consequence of climate change for Russia—the opening of the Arctic Ocean to sea traffic to Asia.

Table 1. Possible path of Russian exports by 2050 (in billions of 2019 US dollars)

	2019	2050
Total exports	424.6	232
Energy exports		
Fossil fuels		
Oil (crude and products)	188.3	75
Gas (pipeline)	41.6	20
LNG	7.9	20
Coal and coke	16.6	5
Nuclear power	5.0	10
Non-energy exports		
Agricultural	8.6	20
Metals	28.6	30
Manufactured goods		
Weapons	15.0	15
Machinery	27.7	30
Gas-driven chemicals	0	7

Data source for 2019: Federal Customs Service, http://customs.gov.ru/statistic/.

But the main impact of climate change on Russia will be external. Russia's economy, as already noted, is based overwhelmingly on the production and exports of natural resources, chiefly energy. There will be two sorts of consequences. First, as governments around the world struggle to deal with carbon emissions, this will affect Russia's energy exports disproportionately. Second, because Russia's main markets for energy exports are Europe and China, any constraints there on fossil fuel imports (whether on prices or volumes, in the form of carbon import taxes) will hit Russia especially hard. Yet Russia has no ready alternatives to make up for a decline in revenues from fossil fuel exports.

If these propositions come true, how much could Russia's export revenues decline? Table 1 distills the major conclusions from the coming chapters, to show how Russia's export revenues in 2050 might compare with today's. It is no more than a sketch—not a prediction—but it underscores the crucial role of oil, in contrast to the more modest potential revenues from other exports.

It is only in the past few years that the full implications of this picture, both internal and external, have begun to penetrate the Russian business and political class. The result is increasing political polarization, as Russian elites debate what to do, and each sector responds in its own way. We turn first, then, to the politics.

1

THE POLITICS OF
CLIMATE CHANGE IN RUSSIA

Climate change is rapidly emerging as a political issue in Russia. It has been a long time coming. Russian climate scientists first began calling attention to the dangers of global warming more than forty years ago, but their warnings were largely disregarded. In the last five years or so, however, awareness of climate change has increasingly penetrated official circles; ministries and companies have formed climate teams and are issuing position papers; politicians and business leaders have begun to take sides in increasingly intense debates; and the media and the public have taken notice. Climate change, one might say, has come of age, at least in Moscow, if much less so in the rest of the country. What has not yet happened is any real action.

Today's Russia, though increasingly authoritarian, is far from monolithic. Behind the façade of one-man rule, it is an active political scene where individuals and groups jostle for preference and influence. The evolving debate over climate change is a prime example. In this chapter and throughout this book, we shall meet some of the major players. Some are climate scientists; most are not. They range from reformers, for whom climate change is an opportunity to promote a broader agenda of economic reform, to conservatives, who reject the view that climate change is a threat to the established model of natural-resource production and export. Some are insider-courtiers, close to the seat of power in the Kremlin; others are outsiders who see climate change as a means to influence policy or promote themselves. Some are business and policy entrepreneurs. Still others are advisors in consultancies and think tanks, which are increasingly present in Russia. To a striking degree, these players' competing views are being expressed in public. Thanks to the internet, the game has become wide open.

These players can be grouped, broadly, into four categories. The first is the community of climate scientists and specialists in related disciplines, such as soil science. This group first launched the study of climate change in Russia, beginning as far back as the 1970s, and it has continued to play an important role in publicizing the direct impact of climate change, focusing particularly on the Arctic. This group includes scientific institutes and some specialized government agencies, such as the Russian Hydrometeorological Service (Rosgidrometsluzhba, commonly called Rosgidromet), which is part of the Ministry of Natural Resources. Russian climate scientists continue to sound the alarm through regular studies and media commentaries[1] while at the same time providing a reality check on some of the more extreme narratives, such as "Siberian methane bombs."[2]

The second is the governmental bodies that are responsible for managing external diplomacy and public relations as they relate to climate change. Governments all over the world are under increasing pressure to develop positions and make climate commitments while defending existing ones. Managing this process in Russia is the job of a growing cadre of governmental climate specialists and consultants who develop strategies and doctrines. Russian climate diplomacy has become active in recent years, as evidenced symbolically by the fact that international conferences on climate change, such as the 2019 United Nations Climate Change Conference in Madrid, now feature official Russian pavilions showing that Russia, too, is doing its part to combat climate change. But this role has become more than symbolic: several ministries, notably the Ministry of Economic Development and the Ministry of Energy, have become campaigners for a more proactive climate policy. Climate change, in short, is becoming institutionalized in the Russian government.

The third group consists of companies that are either publicly listed on foreign exchanges or have sizable assets outside Russia. These companies, some of which are private and others partly state-owned, face growing external threats to their businesses, ranging from the prospect of carbon export taxes in Europe, to pressures from increasingly active foreign investors and regulators, which raise the danger of delistings on foreign exchanges or selloffs of their shares. Many Russian companies now post regular sustainability statements on their websites. Gazprom and Novatek, Russia's two premier gas producers, promote the green virtues of natural gas, while Gazprom is talking about green hydrogen. Rosneft, Russia's state-owned oil and gas company, is planting trees—a million of them in 2019 alone—while showcasing its efforts to reduce fugitive emissions of methane.[3] Rosatom, Russia's nuclear power agency, which has an active international business, advertises the greenness of nuclear power and has developed sidelines such as wind power and waste-to-energy projects. Greenness has become a corporate fashion.

Finally, a fourth group, which one might call the conservatives, opposes any climate-related policies that might limit their freedom of action or impose burdensome reporting requirements. These are a mixed group, ranging from those who still deny that climate change is real, to those who reject that it has negative implications for Russia, whether inside or outside the country. Sergei Ivanov, a longtime intimate of Vladimir Putin who is now Putin's "envoy" for transportation and the environment, remains unconvinced that climate change is man-made. Igor Sechin, the chairman of Rosneft and an equally close associate of the president, denounces as a Western plot aimed at Russia the notion that oil demand could peak. Conservative opinions, particularly from the coal and metals industries, are influential in the Russian parliament, as well as in the leading business lobby group, the Russian Union of Industrialists and Entrepreneurs (RSPP). Conservative opposition from industry has blocked significant action on climate change, such as proposals for a carbon tax or a cap-and-trade market in carbon emissions. This is by far the largest group, and it includes the fossil fuel industries and the ministries that represent industry overall, as well as President Putin himself. This group has a political dominance that is unshaken so far, and it has blocked any meaningful response to climate change by the Russian government beyond the level of rhetoric.

The implications of global trends are conveyed into Russia by a growing number of consultancies and think tanks.[4] The prime example is the Skolkovo School of Management's Energy Center, which plays a key role as a channel of communication between Western experts and Russian policymakers, as it disseminates the latest global developments on climate change through seminars featuring both Russian and international speakers. They are not players, in the same sense as the four groups just named; perhaps they are best thought of as transmission belts, acting as consultants to the companies and the ministries. They have played a key role in raising the overall level of awareness of climate issues in the government and Russian companies.[5]

One group is missing from this list—the public, and public opinion. The Russian public is quick to react to passing environmental events, such as fires, heat waves, and the like. But surveys show that public opinion remains unconvinced by the larger climate-change story, despite the efforts of various nongovernmental groups to publicize it. Regional leaders in the Arctic regularly sound warnings about the consequences of melting permafrost, but their voices remain unheard at the national level. There is as yet no broad public pressure on the government to adopt a more active policy of adaptation to climate change inside the country.

The result is that the politics of climate change remains largely confined to an elite group of government, business, and scientific figures, located mainly in Moscow and to a lesser extent in Saint Petersburg. Of the four groups, only the first is concentrated on the direct long-term consequences of climate change inside Russia, chiefly melting permafrost in the Arctic. The other three, which include most of those who control policy in the government and in business, are focused mainly on developing defensive tactical responses to the emerging challenges arising outside of Russia—managing the international diplomacy of climate change, dealing with the financial threats to Russian companies, and heading off possible climate-related sanctions on Russian exports. For the latter groups, climate change itself remains a remote threat; for them, the immediate issue is managing the growing political pressures on Russia arising out of the global energy transition, and especially Russia's trading partners in Europe.

In tracking the politics of climate change, we will meet some individual personalities that are worth paying attention to. The role of the two main fixtures on the conservative side, Igor Sechin and Sergei Ivanov, has already been noted. To them one should add Nikolai Patrushev, another longtime associate of president Putin who formerly headed the Federal Security Service and is now the secretary of the Security Council, which has overseen the preparation of several of the government's key documents on climate change. These men symbolize the continuing grip of the past.

But more significant as a signpost to the future is the arrival of a new generation of ministers, all around the age of forty, who focus on climate change and the Arctic. A rising young politician, Maksim Reshetnikov, at the head of the Ministry of Economic Development, has led the increasingly active role of that ministry in promoting tighter emissions controls.[6] The recent appointment of the even younger Aleksandr Kozlov as the new minister of natural resources brings to that post a native of the Russian Far East with a strong focus on the Arctic.[7] He in turn has been succeeded at the Ministry of the Far East and the Arctic by the forty-year-old Aleksey Chekunkov, a specialist in investment in the Far East.[8] Their deputies are even younger: the youngest of the lot are the thirty-five-year-old deputy energy minister Pavel Sorokin, born in 1985, and Aleksandr Krutikov, born in 1987, who until recently was first deputy minister for the Far East and the Arctic.[9] These are all active and entrepreneurial personalities, part of an effort by Putin to bring new blood into the government. They are having the combined effect of increasing the government's attention to energy transition and climate change, particularly in the Far North. However, this has not yet produced a coordinated policy; indeed, the new blood may be aggravating the tendency for each ministry do its own thing.

The recent promotion of Aleksandr Novak, the former minister of energy, to the position of deputy prime minister responsible for both energy and environment, places an able and experienced hand in a top spot. [10] As energy minister, Novak earned a reputation as a skilled manager and diplomat, notably as Russia's lead negotiator in the difficult talks with Saudi Arabia over limiting oil exports within the framework of OPEC+, the recently created alliance of Russia and OPEC. In that role he apparently gained Putin's notice and respect. Novak is also known as a reform-oriented financial expert, having served as deputy finance minister under Putin's longtime finance minister Aleksey Kudrin, where he first became acquainted with now–prime minister Mikhail Mishustin. Novak has far and away the greatest personal knowledge of the promise and problems of the Far North, having grown up in the "nickel capital" of Norilsk and risen to senior positions in that industry; by age thirty-six he was deputy governor of the surrounding province of Krasnoyarsk before being promoted to Moscow the following year. Novak appears to occupy the middle ground in Russia's evolving climate change policy. On the one hand, he echoes Russia's traditional defense of its hydrocarbon resources and exports, especially natural gas and liquefied natural gas (LNG), but also coal. Yet he also listens closely to arguments based on climate change, and he appears to accept the "peak oil demand" narrative.[11] He supports prospective energy technologies such as hydrogen and renewables, and in his new position as deputy prime minister he has intervened to push back proposals to cut subsidies for renewables.[12]

It is not clear at this stage who will play the most influential role in Russia's climate policy in the years ahead.[13] As deputy prime minister Novak has the most senior voice, but there are others. When he returned to the Kremlin in 2012, Putin inherited as his climate advisor a noted expert, Aleksandr Bedritsky, who had been appointed by President Medvedev. Bedritsky became an outspoken advocate of strong climate change policies such as a carbon market, but it is reported that

Putin never actually met with him. In 2018, when Bedritsky reached the retirement age of seventy, Putin replaced him with an unknown named Ruslan Edel'geriev. A former prime minister of the North Caucasus republic of Chechnya, Edel'geriev brought to his new job a background in law enforcement and agriculture in the Chechen Republic, but none in climate matters and no experience on the national stage. Edel'geriev's main qualification appeared to be that he was a protégé of the strongman of Chechnya, Ramzan Kadyrov, a favorite regional ally of Putin. His appointment hardly suggested that the position of climate advisor ranked high among Putin's priorities.[14] Edel'geriev's main function appears to consist of coordinating the official climate positions of the government through regular meetings of the governmental Interagency Working Group on Climate Change and Sustainable Development—which meets monthly, and which Putin does not attend—and then representing those positions in international meetings, at which Edel'geriev is the official face of Russian climate policy.

Remarkably, however, Edel'geriev, after an initial two years of nearly complete silence, has rather abruptly begun speaking out in public in favor of strong climate-related measures such as creating a carbon market. (He has also met privately with Russian climate-change activists.[15]) But what may be driving this change of public stance, and whether Edel'geriev has any more direct access to Putin or actual influence than his predecessor did, is uncertain.[16]

Another recent development is the appointment in December 2020 of one of the best-known of the Russian liberal reformers, Anatoly Chubais, as special advisor to Putin on climate matters. Chubais was until then the head of Rusnano, a state-owned company created to promote high-tech innovation. In that capacity he played a major role in promoting renewables in Russia, as described in Chapter 5. But Chubais lost his job when Rusnano was folded into a newly renamed and expanded high-tech venture capital fund called VEB.RF,[17] and his appointment as presidential advisor—reportedly unpaid and unstaffed— may be no more than a consolation prize or a decoration for international consumption.[18] It is perhaps significant that when President

Joe Biden's newly appointed special "tsar" for climate matters, John Kerry, held his first official meeting with a Russian counterpart to discuss climate, that person was Edel'geriev, not Chubais.[19]

In the middle of this swirling mix is President Putin. His public views have evolved over time toward greater acknowledgment of the reality of climate change, but he rejects any dire near-term implications for Russia. His real views are difficult to track, because, as we shall see, he uses two different languages on climate change. Before international audiences, where he can hardly avoid the subject, he speaks about climate change often and occasionally extemporaneously. Before domestic audiences, in contrast, he hardly ever mentions it, except to intervene occasionally in specific environmental episodes such as spills or fires or trash removal. Putin appeared to face a dilemma: on the one hand, he wanted to assure the world that Russia is a responsible climate citizen; on the other, he wanted to prevent climate change from becoming a catalyst for domestic political opposition. It was only in late April 2021, in his annual "State of the Federation" address (called "Poslanie" in Russian), that Putin finally crossed a political rubicon, so to speak, before the Federal Assembly and a nationwide television audience, calling for a vigorous Russian response to the threat of climate change, although as yet with no specific measures. (I explore the apparent reasons for this shift later on in the chapter.)[20] Yet this public ambiguity should not deceive: Putin presides over a system that depends on Russia's hydrocarbon exports. He is clearly aligned with the conservatives, and his modest actions on climate to date set the limits for Russia's actual policies.

To illustrate these themes, this chapter offers a brief historical overview, followed by an analysis of the evolving positions of the players.

The Early Role of Soviet Climate Science

The modern history of climate change begins in the late 1950s and early 1960s, when scientists in both the United States and the Soviet Union observed rising CO_2 concentrations in the atmosphere. As early as

1965, American scientists alerted the White House of the possible harmful effects of global warming caused by the greenhouse effect. Their warnings were based on the meticulous measurements of CO_2 concentrations by Charles Keeling at Mauna Loa in Hawaii, starting in the 1950s, which clinched the case for the steady rise in CO_2 emissions, in what is now famously known as Keeling's Curve.[21]

Their Soviet colleagues were not far behind; indeed, in some respects they were already ahead. Geography has traditionally been a strong area of scientific research in Russia, including such fields as climatology, hydrology, and geomorphology, with research institutes scattered across the country.[22] One of the first Russian scientists to speculate on the possibility that the greenhouse effect would lead to global warming was the climatologist Mikhail Budyko, who began writing on the subject in the early 1960s. Budyko was the first to develop a numerical climate model, which linked the growing atmospheric concentration of CO_2 to human combustion of fossil fuels. The first Soviet conference on climate change and its possible manmade (anthropogenic) origins took place in Leningrad (today's Saint Petersburg) in April 1961. A decade later, in 1972, Budyko published a monograph, *The Influence of Humankind on Climate*. He also publicized his ideas widely in the Soviet popular press, sharing the idea of climate change with a wider readership.[23]

In the 1970s, taking advantage of the East–West détente that prevailed during most of that decade, Soviet climatologists came into close touch with their counterparts in the West. By that time Soviet geographers had measured the steady warming trend in Russia, which has averaged 0.43 degrees Celsius per decade ever since—about twice the global average. The rise was particularly pronounced in Russia's northern latitudes. In the 1980s, when the Soviet Union began opening up under Mikhail Gorbachev, Soviet and Western geographers jointly created the Intergovernmental Panel on Climate Change (IPCC), which was explicitly devoted to bringing the results of scientific research to a wider public and to as well as decision makers, and has played a prominent role in climate change issues ever since. Thus, by the end of the 1980s Soviet climatologists had been active participants in the international climate community for two decades.

By this time, however, more immediate concerns had overtaken the still-obscure debate over climate change. At the end of 1991 the Soviet Union disintegrated, and the political and economic upheavals that followed weakened the Russian state and economy over the next fifteen years. During this period much state funding for climatological research in Russia disappeared, and progress in climate studies came to a halt. Yet many Russian scientists forged ahead on their own, taking advantage of Russia's increased openness to the outside world to strengthen their professional ties with Western colleagues and participate in the growing climate-change community.[24]

The Kyoto Protocol: Russia Enters
the International Arena

The international conference on climate change that convened in Kyoto in 1997 marked a major change in the politics of climate change worldwide. Up to that time, climate change had been mainly a subject of discussion within a narrow circle of scientists and a few officials. At Kyoto and after, it became a global political issue.

The conference had been convened to set, for the first time, mandatory targets and mechanisms for greenhouse gas reductions. But in its circus-like atmosphere, the conference degenerated into a face-off between developed and developing nations, and between Europeans and Americans. The developed nations wanted emission limits to be binding on all; the developing ones adamantly refused this. The Europeans wanted deep cuts in CO_2 emissions; the Americans would not hear of them. The resulting protocol contained so many flaws that the Clinton administration did not even submit it to the Senate, knowing that it would not be ratified, and the Bush administration buried it. By 2004, fifty-four countries had ratified the Kyoto Protocol, but it required one more vote to go into effect. Russia, which had also not yet ratified, now faced a decision. How would it vote?

By this time the international visibility of the Kyoto conference and the subsequent politics of ratification had drawn in policymakers in the Russian government. The prospect of clean-energy projects arising out of the agreement stimulated competition among ministries, especially the Ministry of Economic Development and Trade and the Ministry of Natural Resources.[25]

President Putin, up to that time, had taken no particular position on climate change. His occasional offhand remarks on the subject suggested he thought that for Russia the effects of climate change were more positive than not. (One of his early responses was to joke that climate change would allow Russians to "spend less on fur coats.") The tone of the early Putin Kremlin was highly skeptical. His senior economic advisor at this time, Andrey Illarionov, was a noted climate denier who conducted a one-man campaign against the concept of climate change in both the Russian and international media. He famously denounced the Kyoto Protocol as "totalitarian," "the gulag," and a "global Auschwitz."[26]

However, in 2004, somewhat unexpectedly, Putin decided to support ratification of the Kyoto Protocol. Putin was no more of a convert than before, but he used Kyoto as a diplomatic bargaining chip, trading Russia's ratification of the treaty for the European Union's endorsement of Russia's membership in the World Trade Organization, and he perceived that Russian prestige would benefit from rescuing the treaty at the eleventh hour after the United States had rejected it.[27]

Russia could sign the Kyoto Protocol without fear of repercussions on its economy because the emissions benchmark it assigned itself was based on the Soviet-era level of 1990, when its emissions were very high. Since then, however, with the decline of Russian industrial production after the Soviet collapse, Russian emissions had dropped sharply. In other words, Russia could be a good citizen without actually doing anything during the first decades, while its industrial production gradually recovered and its emissions grew back toward the 1990 level. (That is still the Russian government's position today, as we shall see.)

For the remainder of Putin's second term (2005–2008), climate change went largely unmentioned in the Kremlin. For example, in meetings with German chancellor Gerhard Schröder and French president Jacques Chirac, or with a group of American high-tech entrepreneurs, there was no mention of climate change at all.[28] The same was true in Putin's speeches to domestic audiences, such as his address to the State Council in December 2005 on realizing the highest-priority national projects.[29] Climate change was not one of them. And so it went for the remainder of Putin's second term, until he turned over his president's chair to Dmitry Medvedev—on loan, as it turned out—at the end of 2008.

An Underrecognized Presidency: Dmitry Medvedev (2008–2012)

Dmitry Medvedev, a protégé of Putin from Saint Petersburg days, became president under an arrangement in which Medvedev held the presidential chair, while Putin sat out four years as prime minister, although he continued to exercise power on many issues from behind the scenes.[30] Medvedev, to put it mildly, will be remembered more for his initiatives than for his achievements. That said, during his four years as president he brought a significant change in tone on climate change, consistent with his reformist views on many aspects of public policy. For Medvedev climate change was an opportunity to pursue the modernization of the Russian economy, by promoting energy efficiency (which would automatically reduce greenhouse gas emissions) and renewable energies, although by this Medvedev still understood mainly nuclear power.[31]

It was during Medvedev's term that Russia adopted a number of significant policy documents, including a "Climate Doctrine" in 2009. Medvedev communicated his efforts on climate through frequent video blogs and social media. As with all of his policy initiatives, however, most of this proved to be limited to good intentions and only weak follow-through. For example, his 2009 decree on energy efficiency, which aimed to reduce the energy intensity of the economy by 40 percent by 2020, remained mostly a dead letter.[32]

However, during this period the views of the climatologists were increasingly accepted by official Moscow. For example, the Rosgidromet, which a few years earlier had appeared to dismiss the influence of human activity,[33] in 2008 published its first comprehensive overview of climate change in Russia and its consequences, summarizing the state of Russian research and endorsing the anthropogenic view of the IPCC.[34] This then served as the basis for Russia's first major policy statement on climate change, the Climate Doctrine of 2009, which was endorsed by president Medvedev and has been Russia's official policy ever since.[35]

The Climate Doctrine did not mince words. "Climate change is one of the most important international problems of the 21st Century," it began, and it accepted from the start the anthropogenic origins of global warming, acknowledging that man-made emissions were the principal cause of greenhouse gas emissions. The consequences of climate change were overwhelmingly negative, it wrote, both for the world and for Russia. However, despite its significance as a pathbreaking document, the Doctrine lacked any specific policy prescriptions, and much of the text consisted of general guidelines for equipping the state to monitor and debate the issues.

Under Medvedev the Kremlin began institutionalizing the handling of climate change issues, mainly with a view to developing the Russian government's positions in international forums and bodies such as the G20. Medvedev appointed a respected climatologist, Aleksandr Bedritsky, the head of Rosgidromet, to be his advisor on climate affairs.[36] Bedritsky proved to be an outspoken public advocate, giving frequent media interviews on the dangers of climate change for Russia. Despite this, most of the conversation on climate change during the Medvedev period remained inside individual ministries and public bodies, each one with its own small environmental team, with no particular communication across them. Climate change was on the official agenda, but still only at the periphery of concerns. The prevailing attitude of the Presidential Administration and of official Moscow was "climate pragmatic," in which climate change was seen as an opportunity to push economic efficiency and the "ecological quality" of Russian exports, but without fundamental changes in policy.[37]

During the Medvedev years "climate change" also became part of the vocabulary of Russian industry. Companies were required to report their greenhouse gas emissions or pollutants annually; and, as noted earlier, they began issuing "sustainability reports" on their websites. But for the most part these consisted of ritual phrases coupled with calls for greater efficiency. (Interestingly, though, industrial companies appeared to accept the anthropogenic origin of climate change without much question.) The Russian oil and gas companies, for example, all began referring routinely to climate change in corporate releases, but of all the oil companies, only LUKOIL took an explicit position in favor of international action, reflecting its greater involvement in overseas projects. The notion that climate change might pose some existential risks for the oil and gas industry was not addressed at all and was probably not yet even present in its leaders' minds at this time.[38]

On one oil-related environmental issue, however, the Russian government did take a strong stand. In 2005 it began imposing stiff fines on oil companies that flared "associated gas"—that is, the gas that comes up from wells during oil production. This move was initially motivated by a shortage of gas at the time.[39] But under Medvedev the government strengthened the anti-flaring rules, and it imposed fines on those exceeding the limits.[40] As the flaring of associated gas comes under growing international scrutiny, Russia has joined an international group sponsored by the World Bank, the Global Gas Flaring Reduction Partnership, dedicated to monitoring and reducing flaring worldwide. The policy has had a positive impact, although Russia continues to be the world's largest flarer of associated gas.[41]

In sum, the Medvedev presidency, with the 2008 Rosgidromet report and the 2009 Climate Doctrine, marked the first clear acknowledgment of the reality of man-made climate change and its possible implications for Russia. For the first time we see the clear emergence of the interest groups described at the beginning of this chapter, as the government and the business sector begin to organize to respond to the climate issue. But the Medvedev years also witness the first signs of the split that has come to characterize the Russian response: whereas the scientific and environmental communities (through such documents as the Rosgidromet report) focus on the direct environmental implications and the needed counteractions inside Russia, the other players—the companies, most of the ministries, and the consultancies—concentrate their attention on developing defensive and tactical responses to the diplomatic and financial pressures arising outside Russia.

Putin Returns: The Paris Agreement and Beyond

In 2012, after a largely symbolic election, Putin returned to the presidency and Medvedev became prime minister. The subject of climate change noticeably receded to the sidelines, giving way to more immediate political issues—in particular, the Ukrainian crisis, the annexation of Crimea, and the Russian-supported secession in the Donbass, followed by the imposition of Western sanctions. Forest fires, floods, crop failures, and heat waves made periodic headlines, but without influencing Russian policy or public opinion in any fundamental way.

However, three developments contributed to a new tone in the Russian discourse on climate change. The first was advances in technology, the second was the evolution of climate change as a global issue, and the third was a sharp deterioration in Russia's relations with the West, especially with the United States. The combination of these three changes began to alter the conversation about climate change among Russian policymakers and in the media, including Putin's own public views.

First, the decade of the 2010s was marked by rapid global progress in renewables, mainly solar and wind. In 2010 these could still be dismissed as immature technologies, unable to compete with fossil fuels without state subsidies. Countries that bet on them early, notably Germany, paid a high price for pioneering technologies that were not yet cost-effective without subsidies. But throughout the 2010s, the costs of solar and wind fell rapidly. By the end of the decade, power from renewables had become cheaper than power from fossil fuels, and state subsidies were being phased out in favor of auctions.[42] Solar and wind were becoming fully competitive, even with gas-fired power.

The other major technological trend during this decade was the rise of "tight oil" (sometimes called shale oil), which together with the continuing boom in shale gas suddenly made the United States the leading hydrocarbon producer in the world, surpassing both Russia and Saudi Arabia for the first time since the 1970s.

As the combined result of renewables and the oil boom, there came a major change in the global climate discourse, with the prospect, though as yet highly speculative, that renewables could lead to a decline in oil demand, while expanding supply could produce lower oil prices. The implications for Russia were alarming. Russian institutes and consultancies began spreading the message, and the "peak oil" narrative was soon reflected in Russian policy papers and reports. The conservative camp, including the president himself, remained in denial, but the striking thing is that the peak-oil scenario was quickly acknowledged inside Russian companies and key ministries such as the Ministry of Energy and the Ministry of Economic Development.

The third development following Putin's return to the presidency was a deterioration in the Kremlin's relations with the West. Putin's views on the West and especially the United States, which had already turned sharply negative following the Ukrainian Orange Revolution of 2004 and the wave of "colored revolutions" that followed it, plummeted further after his return to power in 2013 and events in Ukraine in 2013–2014. This carried over into Putin's discourse on climate change, as he began accusing the United States of taking unfair advantage of climate change for self-serving ends.[43]

Yet with Putin back in the Kremlin, the process of institutionalizing—perhaps one should say bureaucratizing—the Russian government's policy machinery on climate change continued. In December 2012 Putin signed a decree creating an Interagency Working Group on Climate Change and Sustainable Development.[44] It brought together ministries and public bodies, the parliament, scientific institutions, and some state-owned companies such as Sberbank, as well as representatives of private business. The Working Group was initiated in response to criticism by Russian companies that there was no place where interest groups, particularly from industry, could meet to make their views heard. Over the last decade the Working Group has ballooned up to an unwieldy fifty members, and most of its actual work now takes place in "expert groups," devoted to specific matters such as preparing the president's speech on the ratification of the Paris Agreement. It is not a significant force for change.

Meanwhile, the government apparatus became increasingly busy producing plans and strategies. Two things are noteworthy about this flow of documents. The first is the growing official recognition of climate change as a threat. Thus the Strategy for Environmental Security to 2025, adopted in 2017, includes climate change for the first time as one of the causes of Russia's environmental problems, and it mandates the creation of a system to monitor the level of greenhouse gas emissions.[45] Since then the focus on climate change has increased further, notably with the publication of a Climate Change Adaptation Plan in January 2020. The national plan focuses not only on the negative consequences of climate change, but also on the potential positives. These include "a reduction in energy consumption for heating, an expansion in crop and livestock production, and an increase in the productivity of boreal forests."[46] Yet in this document, as in others, the references to positive benefits are largely perfunctory, and most of the emphasis is on the negatives.

The second significant fact is a heightened focus on the Arctic.[47] In late 2019 and in early 2020 the government published in quick succession three major documents laying out the government's Arctic strategy. The timing of these documents, if one wishes to be cynical, suggests that they are designed to prepare the ground for Russia's chairmanship of the Arctic Economic Council, an international body. In that respect they are part of Russia's diplomatic effort to promote its image as a good citizen. In addition, they reflect the Kremlin's high priority for the Northern Sea Route (discussed in detail in Chapter 8). But the three documents are also noteworthy for their focus on the domestic socioeconomic repercussions of climate change in the Arctic, particularly as the result of melting permafrost. President Putin himself has reportedly grown increasingly concerned about this problem, following several trips to the Far North, although he continues to downplay it before domestic audiences. This takes us to Putin's complex role in the climate change debate.

The Two Putins

Putin's views on climate change appear to have slowly evolved over time. He now stresses its negative aspects over its positive ones. But overall he remains a conservative voice on climate matters. He insists that Russia is already playing its part in international action to control emissions, and need not do more. He continues to downplay the negative consequences of climate change for Russia, and his specific prescriptions for domestic policy are mainly limited to traditional environmental issues such as trash removal and local air pollution. It was only in 2019 that Putin acknowledged for the first time that the global role of oil might begin to decline in favor of gas and renewables, but he did not see that happening soon; and as we shall see in Chapter 3, he continues to see a strong future for hydrocarbons in the form of gas, and especially LNG. The narrative of "peak oil demand" has begun to appear in his words, but he continues to resist its implications.[48]

Putin's remarks must be taken with a grain of salt. Virtually all of Putin's comments on climate change are addressed to international groups. In contrast, as noted, Putin hardly speaks of climate change before domestic audiences. Putin's annual address to the Federal

Assembly in February 2019, for example, did not include a single mention of climate change, other than to say that Russian production should be "ecologically pure." Similarly, in Putin's annual televised question-and-answer show *Direct Line,* in which he answers questions live on a wide range of issues, Putin gave practically no attention to environmental issues. In the June 2019 session, which went on for more than four hours, Putin mentioned "climate" only twice in passing, while he gave abundant coverage to the problems posed by trash.[49] Finally, in December 2020, in his annual press conference on the state of the nation—which was combined with the "Direct Line" event that year—Putin, again before a largely domestic audience, addressed only specific "ecological events," such as the Norilsk oil spill (described in Chapter 8), but without relating them to climate change.[50]

This disparity poses a mystery. What explains it? One possibility is that Putin feels little domestic pressure to address the wider topic. As we shall see below, Russian public opinion remains focused largely on conventional environmental issues, such as pollution and waste, rather than the broader issue of climate change. Before international audiences, in contrast, climate change is impossible to avoid, but Putin seeks to strike a tactical balance between recognition of the problem and assurances that Russia is on the job. Behind the political tactics, however, Putin seems to be increasingly of two minds.

The more climate change rises as a global issue, the more ambivalent Putin seems to become. This uncomfortable mix of feelings was fully on display in a major address before an international audience in Yekaterinburg in 2019.[51] It featured several themes Putin had never mentioned before in public. For the first time, Putin conceded that the growing problems of climate change—droughts, crop failures, natural disasters—were anthropogenic in origin. Also for the first time, Putin put the consequences of climate change for Russia squarely at the center of his speech. "In Russia these effects are felt especially sharply," he said; "temperatures in Russia are rising two-and-a-half times as fast as the planet as a whole."

But at this point in the speech, Putin launched into a sharp attack on the state of the global dialogue on climate change, and especially renewables:

> Instead of reasoned discussion we see populism, speculation, and . . . obscurantism. It gets to the point where there are calls for the world to reject progress, which will allow at best a preservation of the status quo, the creation of local welfare for the chosen few. . . . Such an archaic approach is a path to nowhere, to new conflicts . . . to a migration crisis, in Europe and the United States.

Yet at virtually the same moment the Duma passed, by an overwhelming majority, the first reading of a new law on climate. A more minimal measure would be hard to imagine. For example, it requires heavy polluters (above 150,000 tons of CO_2 per year) to begin issuing reports on their emissions—but it provides no mechanism for verification. Companies will be "allowed to develop projects to cut their carbon footprint," but on a purely voluntary basis. In the debate on the floor of the Duma, there were still claims by some deputies that climate change was an invention of hostile Western interests, and that it was the "scam of the century." The draft climate law had actually been under discussion since 2017, but was stalemated for four years by conservative opposition, mainly from industry. Earlier versions, from the Ministry of Economic Development, had included emissions quotas, emissions trading, and penalties for above-quota emissions. But all of these provisions were dropped in the new draft climate law, leaving only a vague reference to tradable carbon credits. In short, on balance the Russian government and most of industry remain resolutely opposed to meaningful action to reduce emissions.[52]

As the prime example of the wrong approach, Putin cited the rejection of nuclear power and hydrocarbon fuels, and the "absolutization" of renewables, which he denounced as "sweeping the trash under the rug." While conceding that "wind power is good," he went on to describe the negative environmental effects: "Does anybody remember about the birds? How many birds die because of windmills? They vibrate so much that earthworms crawl out of the ground."[53] How comfortable will we be, he wondered, when the entire planet is covered with windmills and solar panels?[54]

In another international conference, in April 2019, Putin asserted that the relative shares of energy sources in the global economy would remain roughly unchanged in the future. "We do not see a transition. So far there is no critical transition from hydrocarbons to renewables, . . . 'Critical,' that is, from the standpoint of those who produce oil, gas, and coal." He conceded that there might be lower oil prices ahead, but he insisted that Russian oil would remain profitable at $30 to $35 per barrel. "I don't see any threats here in overall perspective. . . . I simply don't see any threats to us, they don't exist."[55]

As alternatives, Putin went on, there are technologies that are "similar to natural processes." One of them is nuclear fusion, "which produces light and heat in the same way as our star, the Sun." Thus, he argued, the only way forward is through more high technology, to supply the clean power and other solutions that will be required. Progress and growth remain the supreme good.

Despite these strong words, Putin increasingly recognizes the need for defensive action, at least where Russia's dependence on oil is concerned. At a meeting with international investors in October 2020, Putin asserted that Russia had already taken the necessary fiscal measures to lessen the dependence of the budget on oil revenues. "The structure of demand will change," he said, "but it will change slowly."[56]

By the fall of 2020, Putin's conversion to the climate change story appeared to have finally become complete—but again, only before an international audience. In an address to the annual Valdai conference, Putin gave his most graphic description yet of the "extremely negative" consequences facing the Russian economy and the population, particularly from melting permafrost. Not only does melting permafrost threaten pipelines and buildings, he said, but it threatens to release massive quantities of methane, thus setting off a positive feedback effect—the warmer the temperature, the more methane—that could spiral out of control, setting the earth on a course that would eventually make it as hot as Venus.[57]

Where do Putin's ideas on climate change come from? Not, apparently, from climate scientists themselves. Rather more significant as a possible influence on Putin's thinking is Sergei Ivanov, a longtime ally of Putin from Saint Petersburg days whom Putin once described as an "elbow partner." After serving in a variety of leading roles throughout the Putin years, Ivanov is presently a member of the Security Council, and has been named special presidential representative for environment protection, ecology, and transportation. But Ivanov is a climate-change skeptic, and he has shown little interest in environmental matters, except, curiously, for the defense of East Siberian leopards and Pacific dolphins. In a television interview in 2017, Ivanov said, "In order to draw far-reaching conclusions, scientific data are needed for a period of at least 1,000 years. So far, humanity has data only for 100 years. And this is nothing, just nothing." As for Putin's reference to Venus, the source of that is unknown.[58]

On balance, whatever he may say before international audiences, Putin remains a reluctant recruit to the climate-change story. This in turn gives encouragement to the conservative camp. Even within Putin's inner circle, there is strong conservative opposition to anything that might limit Russian companies' freedom of action. This was clearly on view in the struggle that broke out in 2015 over the ratification of the Paris Agreement.

The Battle over the Paris Agreement

Three years into Putin's third term, a major international meeting in Paris in 2015, after much negotiation, produced an international agreement under which the signatories committed themselves to policies that by 2050 would limit the warming of the atmosphere to 2 degrees Celsius above preindustrial levels. The Paris Agreement imposed no new legal obligations and was thus essentially voluntary.

But Russia's participation in the agreement required ratification by the government, and this touched off strong opposition from the Russian business sector. The RSPP campaigned vigorously against ratification, addressing multiple letters to Putin warning of the harm to the Russian economy if any of the measures discussed in Paris—and notably a carbon tax—were implemented in Russia.[59] Confronted with this resistance, Putin hesitated. Although he had spoken out in Paris in favor of the agreement, he postponed ratifying it for more than three years.

Up to about 2019, Russian politicians had taken little public part in debating climate-change issues. But in the spring of that year, as the Duma took up the question of Russia's ratification of the Paris Agreement, conservative members began to speak out against it. In Paris, Putin had supported the final declaration of the Paris conference, and he promised that Russia would ratify it. But when the text of the agreement was submitted to the Duma, it provoked an unexpectedly sharp resistance, in which representatives of industry, led by the head of the opposition, the A Just Russia party faction, Sergei Mironov, vigorously opposed ratification.[60] Ratification was just the tip of the iceberg, said Mironov. The real problem ahead was its coming implementation. He denounced pressure from the Presidential Administration to ratify the accord by the end of the year. Mironov objected to the rush to sign, which allowed no time for public discussion.

In the debate in the Duma, several deputies continued to deny that climate change was man-made. "No more than 5% of greenhouse gases comes from human action," claimed Igor Ananskikh, the deputy head of the Duma's energy committee and also a member of the A Just Russia faction. The Paris Agreement, he said, was clearly an attempt to impose limits on Russian energy production.[61] All the old theories about nonanthropogenic origins were trotted out, notably by nonclimatologists such as Valery Fedorov, of the geography department at Moscow State University, who blamed changes in the earth's orbit. Another denounced the Paris Agreement as a scam (*moshennichestvo*), based on manipulation of the data, and asserted that arctic ice was not retreating but expanding. The consequence of ratification, he said, would be a 10 percent decline in GDP; all branches of the economy would become unprofitable, and 10 million Russians would be thrown out of work. The discussion grew so heated that the representative of the Ministry of Economic Development walked out, leaving the representative of the Ministry of Natural Resources to face the storm alone.

In the end the government withdrew its bill, and it ratified the Paris Agreement by a simple executive decree from then–prime minister Dmitry Medvedev. This procedure was defended as the legal equivalent of ratification, on the grounds that the Paris Agreement was a "framework law" (*ramochnoe soglashenie*) that did not require any change in existing Russian legislation and therefore did not require a vote by the Duma. On technical grounds the government was correct, but opposition to ratification was so great that the government was evidently in some doubt whether it would pass.[62]

In October 2019 the Ministry of Economic Development submitted to the government a draft law, "State Regulation of Greenhouse Emissions," that had clearly been substantially watered down compared to an earlier version prepared in the spring. The initial version followed the model adopted by several European countries: it called for state monitoring of emissions and company-by-company limits on greenhouse gas emissions and fines for exceeding them, and it provided for the creation of a cap-and-trade market for carbon. But the draft had aroused strong opposition from industry, represented by the RSPP and the Ministry of Industry and Trade and the Ministry of Energy. In the final version, the limits and the fines had been deleted, replaced by voluntary quotas, and the carbon market was gone. In addition, a proposed fund to provide financial aid for climate measures in developing countries—a key feature of the Paris Agreement—was eliminated. In effect, as the Russian daily *Kommersant* commented, the weak bill amounted to joining the Paris Agreement without actively participating in it. All the hard parts had been swept under the rug.[63]

A similar fight broke out in the spring of 2020 over another draft plan, also written by the Ministry of Economic Development, that was devoted to the long-term development of the Russian economy to 2050. The ministry proposed in its first draft to put a price on greenhouse gas emissions and to lower Russia's emissions to 67 percent of the 1990 level by 2030.[64] This brought an immediate riposte from the RSPP, which denounced the ministry's "substantial rise in ambitiousness" and the "risk of additional financial burdens for investors." It demanded a change of the 2030 target to 70 percent of the 1990s level, to be confined to voluntary measures only. The battle went on all through the summer. Finally, in November 2020 Putin weighed in on the side of the industrialists, in a decree that endorsed the RSPP's demand for the easier 70 percent target and did not mention the proposed carbon price.[65]

The New Battleground: Carbon Border Taxes

In the past two years, a new development has abruptly raised the visibility and urgency of the climate change issue in the Russian elite. In 2019, as part of its new "Green Deal," the EU Commission proposed levying a special tax on any imports that did not meet the EU's carbon emissions standards. This new measure, called the "carbon border adjustment mechanism" (or CBAM), was designed to create a level playing field between European producers, who pay carbon taxes, and foreign producers, who do not. The CBAM would raise between €5 and €14 billion per year to help finance the EU's climate change program.[66]

Russia was suddenly in the line of fire, and the result was to raise sharply the visibility and urgency of the climate change issue among Russian companies and government policymakers. Until then, even the advocates of stronger measures on climate change had thought that border taxes, if they came at all, would only arise in the distant future, and that Russia would have plenty of time to prepare. But suddenly Russia faced an imminent threat to its prime export market, which in

2019 had brought in $189 billion, or close to half of its total export revenues. Russia was caught unprepared. As Ruslan Edel'geriev noted in a signed article in June 2020, Russia had no mechanisms in place to calculate the carbon emissions of individual exporters, no corporate low-carbon strategies, and no means of setting appropriate mitigation targets.[67] Consultants and think tanks immediately went to work to calculate how much Russian exporters might have to pay the EU. The Anglo-Dutch group KPMG, one of the Big Four accounting firms, came up with three scenarios; in the base case, the border taxes would cost Russia $33 billion between 2025 and 2030.[68] The primary victims would be Russian exporters of gas and copper and nickel; in contrast, Russian oil exports would be relatively unaffected. But other exporters saw themselves as under threat as well; thus the Russian steel industry, which relies principally on coal, claimed it would face losses of as much as $800 million per year from the EU tax. The RSPP was predictably opposed to any and all concessions and quickly mounted the battlements, demanding that the Russian government take a firm stance against any carbon border taxes, including an appeal to the World Trade Organization and lawsuits before European courts. But the previously solid unity of the RSPP has begun to show cracks, as the oil companies distance themselves from the coal and metals companies.[69]

The EU's proposed carbon import tax has stimulated a wide range of counterproposals, ranging from calls to "recount the trees" and increase their claimed absorptive capacity[70] to a new measure to use the remote island of Sakhalin, on Russia's Pacific coast, as a "test bed" to experiment with a carbon trading scheme.[71]

The net effect of the EU's proposed scheme has been to strengthen the voices of those Russians—hitherto a small minority—who have been calling for a carbon tax or a trading scheme. Far better to levy a carbon tax ourselves than to allow the EU to do so, says the World Wildlife Fund's (WWF) Aleksey Kokorin. But so far the Russian government seems to be leaning toward compensating the Russian companies in the form of tax breaks and "preferences." Aleksandr Shirov, of the Russian-European Center for Economic Policy, argues that it would be cheaper to do that than to impose a carbon tax on the entire range of Russian emissions.[72]

The prospect of carbon export taxes appears to have prompted a key event in April 2021: Putin's annual "Poslanie" to the Federation Assembly, which was televised nationwide. In it Putin declared, "We must meet the challenges of climate change, adapt agriculture, industry, housing, our entire infrastructure. We must create a new industry to make use of carbon wastes; we must lower emissions and implement a strict system of oversight and monitoring. Over the next 30 years the cumulative net volume of GHG in Russia must be lower than in Europe."[73]

Meanwhile, it remains far from certain when or how the EU will actually implement a carbon import tax. But the threat alone has put climate change on the agenda of the Russian government as nothing else previously could. Once again, however, it is clear that for most of the Russian players, foreign trade remains a more powerful motivator than climate change. But the threat of a carbon border tax is also reinforcing the position of those, such as Anatoly Chubais, who call for a carbon tax in Russia. This will be a key test of the relative influence of the various Russian elite voices on climate change, as well as Putin's hitherto strong alignment with the conservative side.

Climate Change in Russian Public Opinion

Public opinion has been largely absent from the Russian debates over climate change. Could climate change become the subject of antiestablishment generational protest in Russia, as it abruptly has become in the West? So far there has been little sign of it.[74] The only issues that have brought crowds into the streets are trash removal and occasional opposition to projects that involve cutting through forests and protected areas.

Nongovernmental organizations (NGOs) such as Greenpeace and the WWF are present in Russia, and their views are regularly reported in the press as well as on social media, but it is difficult to know what their audiences are or how much influence they may have. The government has shut down many NGOs in the past decade, especially those with foreign funding, so the fact that environmental NGOs have survived to the present suggests that they may enjoy discrete support from some government circles. The WWF has taken the interesting initiative of conducting an annual review of the "environmental transparency" of Russian energy companies, and the review is widely reported in the media. The companies cooperate, seemingly mainly with an eye on international investors, but the practical impact on domestic policy is doubtful.

Most Russians remain largely uninformed about climate change, although awareness has risen gradually. Russia's one remaining independent public-opinion polling organization, the Levada Center, has been conducting surveys of Russian opinion on climate change since 2007. The most recent poll by the Levada Center, conducted in December 2019, had "environmental pollution" leading all of the respondents' worries by a wide margin, with 48 percent versus 42 percent for "international terrorism." "Climate change" also ranked high on the list, in fourth place, with 34 percent of respondents naming it as their chief concern.[75] But the different scores for the two suggested that for most of the respondents the traditional category of pollution still topped that of climate change, which remained somewhat abstract. Moreover, a comparison of the 2019 results with those of a decade earlier, in 2010, suggests that climate change had not risen much on the list of concerns, whereas the category of utilization of waste had doubled.[76]

A follow-up poll by the Levada Center in September 2020, focused solely on climate change, showed that an overwhelming majority of Russians (93 percent) agreed that there had been changes in the climate, but opinions were sharply divided over the seriousness of the problem, with 52 percent calling it "significant" while 40 percent considered it "overblown and far-fetched." Between two-thirds and three-quarters, however, said that they would be unwilling to pay more for alternative energy sources, and 94 percent supported planting trees instead. Two-thirds believed that the EU's proposed carbon export tax was nothing more than a revenue-raising scheme.[77]

The implication of these findings is that in the absence of broad public focus or concern, there are hardly any domestic political pressures on the leadership over the issue of climate change.[78] There is nothing in Russia like the groundswell of public opinion that has produced zero-carbon pledges in other countries around the world or in individual states in the United States. Climate change has not yet become fused with other political and social issues, as in the "Green New Deal" in the United States, or become a "children's crusade," as in Europe. Protests over environmental issues do arise from time to time, but they are focused on local complaints, such as waste disposal. Climate change as such is not yet a political issue in Russia.[79]

One partial exception to the prevailing public indifference—but one that only underscores the rule—is forest fires. In 2019 there was a serious outbreak in East Siberia and Sakha that burned nearly 20 million hectares (200,000 square kilometers), a new record.[80] For a brief time the fires sparked outrage throughout eastern Russia, which was relayed to Moscow via the internet. An Instagram site called #Sibir' became overnight one of the most popular in Russia. Moscow's entertainment world got involved, as singers and actors launched a flash mob called "Siberia Is Burning."[81]

By late July, official Moscow was in an uproar; there were impassioned debates on the floor of the Federal Council (the upper house of the Russian parliament) over who or what was to blame. Putin sent military units to Siberia, and he dispatched then–prime minister Medvedev to coordinate the effort. But by mid-August the annual rains had begun to arrive in East Siberia, and the fires began to shrink. As they did so, media attention and public interest rapidly waned. Politicians went on to other business, and life returned to normal. The following year, much the same cycle was repeated, although because of the COVID-19 pandemic the reaction of the media and the public was less intense. Putin's responses to the forest fires suggest that he is attentive to the possibility that anger over the fires could combine with other issues to become a sustained popular issue. But so far that has not happened.

A similar episode—a massive oil spill that occurred on the site of Nornikel' at Norilsk in the spring of 2020—demonstrates much the same point. For a few weeks, both the conventional media and social media covered the spill closely. President Putin became personally involved, and the principal owner of Nornikel', Vladimir Potanin, faced reprimands and heavy fines. But as we shall see in Chapter 8 on renewables, the larger issue—namely, the acceleration of melting permafrost as a result of climate change—was hardly ever mentioned, and certainly not by Putin. In short, pollution, rather than climate change, remains the focus of Russian public discourse.

In sum, perceptions of climate change have slowly evolved in Russia over the last two decades, from an abstract question that was initially confined to climatologists to one that is actively debated in government and business circles and in the media, although much less so in public opinion. Yet it is still oil, and secondarily gas, that make the weather in Russian climate politics. As Tat'iana Mitrova, the influential research director of the Skolkovo Energy Center, concludes flatly, "Let us be frank: the Russian economy and the government system as a whole are not ready for decarbonization and the energy transition."[82] At present there is nothing within the country that would cause the Russian leadership and elite groups to change their traditional model, which remains based on the development and export of hydrocarbons. But that model is increasingly vulnerable, primarily from forces outside Russia. In the next chapter we turn to the number one source of vulnerability: oil.

2

THE TWILIGHT OF RUSSIAN OIL?

Igor Sechin, the CEO and chairman of the board of Rosneft, Russia's state-owned oil company, is the most powerful man in the Russian oil industry today.[1] Yet he is not an oil man by background, but a linguist. At Leningrad State University (as it was known then), Sechin majored in Portuguese and French, both of which he speaks fluently. After seeing service in Portuguese Africa in the 1980s, Sechin came back to Leningrad, where he met Vladimir Putin. The story goes that in 1990, when Putin visited Brazil as part of a goodwill delegation of Leningrad officials, Sechin was the interpreter. The two men hit it off, and when they returned home Putin offered Sechin a job. The rest is history. For the next thirty years Igor Sechin was Putin's closest aide and associate, and the bond between them, though strained on occasion, has never been broken.[2] In the late 1990s both men retrained themselves in energy economics, taking *kandidat* degrees (the equivalent of a US PhD) at the Saint Petersburg Mining Institute.[3] Thus, Sechin is knowledgeable about energy matters, and during his years in the Kremlin he was responsible for various aspects of Putin's energy policy.

Sechin is also one of the most powerful voices on Russian climate policy. Sechin's views on oil and climate change are highly conservative, and they broadly coincide with Putin's own positions, as analyzed in Chapter 1. Global oil demand, Sechin asserts, will remain stable for a long time; even though oil's share in primary energy demand may decline, absolute oil consumption will increase by 10 percent by 2040, and by 20 percent in Asia, especially in India, driven by population growth and rising standards of living. A forced "greening" of energy, Sechin warns, will cost the world dearly, and while he calls for a balanced approach, it's clear that for him "balance" means that Rosneft does not need to be proactive in turning away from hydrocarbons in the pursuit of the energy transition agenda. The best defense would be in monetizing its vast reserves; in practical terms, this means full speed ahead for oil.[4]

Sechin's views put him at the opposite pole from an increasingly green trend among the international oil majors, in particular that of Rosneft's partner and minority shareholder, BP. Under its new CEO, Bernard Looney, BP has strongly embraced the "fast transition" energy narrative, and is basing its long-range strategy on a radical exit from hydrocarbons in favor of renewables.[5] Sechin has no such plans, and neither does Putin.

There is, however, an important difference in tactics between the two men, and this could be clearly seen in Russia's decision in 2016 to break with long-standing Russian policy and to form an alliance with Saudi Arabia and its OPEC partners, the so-called OPEC+ group, to limit production and keep oil prices high.[6] Sechin, in contrast, publicly and repeatedly opposed any restraints on Russia's oil production, and called instead for increasing exports, even at the cost of lower oil prices.[7] Despite Sechin's discontent, however, Russia has so far maintained the alliance. We examine it in more detail below.

In this chapter I explore the future of Russian oil and of the Russian oil industry in the age of climate change. This is the single most important question in thinking about the future evolution of Russia as a whole. Oil makes the weather in Russia. But what will the weather—climate change—do to the future of oil?

The Energy Transition—Fast or Slow?

On the eve of the COVID-19 pandemic, the future of oil had become the subject of an increasingly impassioned debate around the world. It was essentially a battle between two narratives. As outlined in the

Introduction, the first was a slow energy transition; the second a much faster one. The slow-transition narrative was the consensus view, its rationale focused primarily on economic drivers of change. On oil specifically, Western energy companies and most consultancies agreed on a basic picture: oil demand, after growing strongly over the next two decades, would peak in the 2030s or early 2040s, and then begin a long but slow decline. Nevertheless, oil and gas together would still account for about half of total energy demand by mid-century. Renewables would increase spectacularly, yet their share of global demand would still be only 15 percent at best. The big winner would be gas (discussed in Chapter 3); by 2050 its share would either overtake that of oil or come close to it. The big loser, all agreed, would be coal.

In this first narrative there were two main messages. The first was that a peak in oil demand was coming—in itself a remarkable change in industry thinking compared to only a decade earlier. But the second message was that the absolute volume of oil consumed by the world at mid-century would be even greater than it is today; only in the following decades would it begin to recede slowly below present levels. Indeed, oil demand and production by 2030 could rise to 114–117 million barrels a day (mbd), compared to about 98 mbd in 2019, for a net increase of nearly 20 mbd.[8] In short, the sun was beginning to set on the age of oil, but it would be a long twilight.[9]

The planning scenarios of the oil companies pointed to two main sources of growth in demand for liquids: transportation and petrochemicals. BP estimated in 2019 that despite gains in efficiency and penetration by electric vehicles, energy demand for transportation, most of it consisting of oil, would increase by 20 percent by 2040; ExxonMobil put the rise at 25 percent. But transportation would no longer be the main driver. Both BP and ExxonMobil projected that the chemicals sector would account for about one-half of the net growth in global demand for liquids. The future of liquids demand would increasingly be plastics.[10]

But a second narrative, in which economic drivers are overshadowed by the alarm over climate change and political ambitions to it, has become increasingly powerful in the last few years, even among some of the oil companies. It holds that the transition away from fossil fuels will happen much faster and sooner than in the first narrative. Fast-transition scenarios point to the powerful forces for change—chiefly technological, but also financial and political—that are driving a rate of adoption of new energy technologies that is unprecedented in human experience.[11] The slow-transition story focuses on the *stock* of present energy structures, and thus on its inertia and resistance to change, whereas the second focuses on the *flow* at the margin, particularly the changing behavior of the players, particularly investors. Not surprisingly, the energy industry tends to favor the former, while financial analysts and the investment sector, including a growing number of institutions such as pension funds, lean toward the latter.[12]

In this debate, climate change is only one of a mix of issues. In Asia, opposition to fossil fuels is driven primarily by concern over pollution and its impact on public health, not by CO_2 emissions or the notion of climate change. In Europe, however, anxiety over climate change is increasingly the strongest driver of energy policy, leading to tougher government policies under the heading of decarbonization. In the United States, key states such as California have also promoted decarbonization, while the federal government retreated under Trump. But the key change has been the attitude of the financial community. Climate change has become part of a blend of causes called "ESG"—environment, social, and governance—which investment institutions and activist shareholders are invoking in an expanding contest of wills with corporate managers. Even now, in the midst of the pandemic, the battle between the first and second narratives is playing out in shareholders' meetings and in corporate boardrooms. Many energy companies are altering their investment strategies, mainly in the direction of the second narrative, although it is difficult to tell how much is genuine conversion and how much is a forced-march response to mounting losses and declining share prices.[13]

What might happen to the two narratives in the years after the COVID-19 pandemic? Both narratives will be weakened, and there is a possible third narrative, which might be called, "slower for longer."

As I write this, the pandemic is on track to wipe out the last ten years of GDP growth. International trade and investment, which had never quite recovered to the levels before the Great Recession of 2008–2009, could well stagnate through the 2020s and perhaps beyond. In the energy sector, the overall effect would be to slow the energy transition away from fossil fuels, especially from oil and coal. The forces that were driving change at the margin (such as investment in electric cars or in renewables) might weaken, while the life of existing stock (for instance, replacement of old cars or investment in new buildings) could be prolonged. Concerns over energy security could increase the tendency to invest in domestic resources. Coal, in particular, might gain a new lease on life. Yet paradoxically, the "slower for longer" narrative could also see lower oil demand, because of the overall slowdown in GDP growth and the general lessening of prosperity, which are the two main drivers in the first narrative. In addition, changes in work habits could cause a decline in oil consumption for transportation, its largest single use. Indeed, the latest oil-company scenarios, particularly in Europe, see the peak in oil demand coming sooner. According to BP, peak oil demand might already be here.[14]

The world has learned to be suspicious of "peak oil" narratives. It was not so long ago that "peak oil" meant "peak oil supply." The global oil market inflated into a bubble of high prices that lasted for a decade, until it burst in 2007–2008 and deflated again in 2013–2014 and 2020–2021. It is a cautionary tale: every one of the "catechisms" that underlay the peak oil supply narrative is present again in the "peak oil demand" story, only in reverse. The world is not running out of oil; it is (supposedly) in permanent oversupply. China will no longer drive the market upward; its growth is slowing down. Saudi Arabia has unreported excess capacity and cannot afford to cut production for long. Above all, tomorrow's world will no longer want as much oil as today, thanks to electric cars and politicians' vows of "net zero" carbon emissions. Tomorrow's oil will remain in the ground.

In this chapter I will not attempt to predict how this swirling mix of predictions will play out. That is a task for a full-scale scenarios exercise, which lies beyond the scope of this book and which would be premature in any case. Rather, my intention here is to focus on how these global narratives interact with Russian oil policy and performance. As we shall see, Russian strategists in companies and ministries, as well as a growing range of Russian think tanks, are aware of these stories and their possible implications for Russia. This growing awareness coincides with an increasingly anxious debate over worrisome trends within the Russian oil sector itself. What does "peak oil demand" mean for Russia?

First, a word about what is at stake. Oil exports are far and away the largest source of Russian export revenues, which in turn are central to the Russian government's income. In 2019, the last "normal" year before the pandemic, income from oil exports, at about $188 billion, accounted for 44 percent of all Russian exports,[15] exceeding other exports by a wide margin. Together with gas exports, which brought in an additional $49.5 billion in that same year, hydrocarbon exports totaled $237.8 billion, or 56 percent of Russia's export income. Oil and gas revenues (including a small share of domestic taxes) made up 39 percent of the federal budget.[16] These numbers are acutely sensitive to world prices. In 2016, for example, when world oil and gas prices plunged, hydrocarbon revenues in the federal budget dropped by two-thirds compared to two years before.[17] In 2020, when prices plunged again, the impact on federal revenues was even more severe.[18]

Therefore, clearly the single most important question for the future of Russia in the era of climate change is the outlook for hydrocarbon production and exports. This chapter tells the story of oil; Chapter 3 takes up that of gas.

Profile of the Russian Oil Industry

Oil was one of the top priorities of the Soviet command economy from the 1960s onward, and by the mid-1980s, on the eve of the Soviet collapse, the USSR was the world's largest producer.[19] But it was also one of the most wasteful and inefficient, and the post-Soviet oil industry suffered from a legacy of damaged fields and destructive practices. These problems, when combined with the chaos of post-Soviet privatization, caused oil production to drop by half in the 1990s. However, this situation also created opportunities, as the Soviet era left a large overhang of discovered but undeveloped reserves. As the country and the industry stabilized in the early 2000s, there followed a strong recovery of oil production, and output continued to grow steadily over the next two decades. In 2018, Russian oil production finally broke the all-time record established in Soviet times, with 556 million tons (mt), or 11.1 mbd.[20]

In the 1990s the Soviet oil industry broke up, and its various pieces were cast loose. They quickly became the object of intense battles between insiders and outsiders. The result was a completely new alignment. The best bits were privatized, under names that briefly became famous at the time, like Yukos, Sibneft, LUKOIL, and Surgut. The Russian government ended up with the least desirable remnants, essentially the lifeboats of the sunken Soviet ship, which it gathered together as a state company under the name of Rosneft. Under Putin there followed more battles, as Putin moved to recentralize the industry under state control. Sibneft was acquired by Gazprom, although it remains a largely autonomous entity under the gas giant. Yukos, following the imprisonment of its CEO Mikhail Khodorkovsky, was absorbed by Rosneft. LUKOIL and Surgut, privatized by Soviet-era insiders, are the largest remaining private companies, together with a handful of mainly regional producers, such as Tatneft, based in Tatarstan. Altogether, Rosneft, led by Sechin and strongly backed by Putin, is by far the dominant company, with 38 percent of total production.[21] Today the Russian oil scene has largely settled down, although there may be further battles ahead, as the founders of the remaining private companies retire, or when Putin finally departs.[22]

However, despite the many changes it has undergone in the past thirty years, the Russian oil industry is still largely based on the infrastructure and resources of the Soviet era, above all the vast oil-fields of West Siberia. Today West Siberia remains the workhorse of the Russian upstream, supplying 57 percent of Russian crude production in 2017.[23]

But the core of West Siberian production has been declining for a decade.[24] The onset of climate change comes at a time when the Russian oil industry is facing the end of the long West Siberian era. But no major new provinces have been discovered to take its place on anything like the same scale, and there have been no fundamental technological breakthroughs, such as unconventional tight oil, that would change the picture. Offshore oil from the Arctic is too costly to invest in at present prices, and in any case Russian industry lacks the technology to explore or develop it. Most of the foreign technology that could tackle the Arctic offshore is currently under Western sanctions, with no end in sight. For the next two decades at least, the oil industry will have no choice but to intensify its effort in West Siberia while doing its best to contain the decline there. But by mid-century West Siberia will be fading rapidly.

Three Colors of Oil

In my earlier book on Russian oil, *Wheel of Fortune*,[25] I distinguished "three colors of oil" as a convenient way of thinking about the architecture of Russian oil. "Brown oil" is oil from existing fields, mainly in West Siberia and secondarily the Volga-Urals and the Timan-Pechora regions. "Green oil" comes from new fields, both on the periphery of established areas and in new provinces such as East Siberia. "Blue oil" is offshore oil, particularly the Arctic deepwater, but also shallow offshore areas along the coastline of the Arctic Ocean.

A decade later this remains a useful shorthand, particularly to track the changes that have taken place in the last ten years. Brown oil comes primarily from so-called mature fields, in which half of the original reserves have already been produced. Many of these were already in production in the Soviet period; as late as 2008 such "legacy" fields still accounted for 60 percent of Russian oil output. But even then they were declining, and the decline has continued, despite the intense efforts of the industry to hold the line. This means higher costs. For example, every year, to maintain well pressure, operators pump into their brownfields the equivalent of a Siberian river, a technique known as "water flooding" (the familiar quip is that the Russian oil companies are the largest water companies in Europe). This implies that they are also some of the major consumers of electricity, which is needed to pump the water. Yet it is an inexorable retreat; every year the so-called water cut (the ratio of water to oil in the liquid that comes up from the well) increases, and the amount of electricity consumed per ton of oil grows steadily. Even so, the Russian brownfields produce some of the cheapest sources of oil in the world outside of Saudi Arabia, and they remain the mainstay of the Russian industry.

The news is at the margin. Most of the net growth of oil production over the past decade has come from the greenfields—new fields developed since the Soviet era. This was particularly striking in 2014–2016, when the oil industry was hit by sanctions and lower oil prices. Oil from greenfields increased by about three-quarters during that period, offsetting the decline in the brownfields.[26] However, most of the increase in greenfield production was the result of investments made prior to 2012, when prices were high, capital was plentiful, and the sanctions had not yet struck.

The investment environment today is much more difficult: oil prices are lower, capital is scarcer, and sanctions are hitting key activities. If anything, these constraints will grow more severe in coming years. The challenge to the Russian oil industry will be twofold: to hold the line in the brownfields while increasing production in the greenfields despite their higher cost. The greenfields will make a modest contribution, but the rearguard battle in the brownfields will determine the overall trend. Meanwhile, there will be little or no contribution from the bluefields.[27]

To what extent do the Russian government and the oil companies share this picture, and how are they responding? As we shall see in the rest of this chapter, Russian oil policy has been paralyzed by controversy. The prime illustration is the tangled history of the draft *Energy Strategy to 2035*, finally approved by the government in June 2020, after a five-year delay caused by multiple disagreements among the ministries and resistance from the oil companies.[28] The document reflected the views of then-minister Aleksandr Novak, a former deputy minister of finance, and those of his deputy Pavel Sorokin, a specialist in international energy finance. Through players such as these—specialists in finance rather than conventional oilmen—new views began to reach the Russian oil world, but they encountered sharp opposition.[29]

A Cry of Alarm: The *Energy Strategy to 2035*

The *Energy Strategy* is a cry of alarm from the Ministry of Energy to the government and the industry. Clearly intended as a wake-up call, the *Energy Strategy* is remarkable for its pessimism. In particular, it accepts the "global peak oil demand" narrative; it projects that growth in oil demand will slow after 2025 and peak before 2030, partly because of the spread of renewables but also because of gains in the efficiency with which oil is consumed in the world. As a result, Russia will face a much more difficult export environment ahead.

But its chief message concerns Russia's own oil production.[30] According to the ministry, Russian oil output faces an acute danger of decline. There are two main problems: a low level of funding for exploration, as a result of which exploration drilling has declined in the last ten years; and inadequate technology, the consequence partly of sanctions but also of the oil industry's long-standing heavy dependence on imported equipment and services.[31] Unless these problems can be addressed, oil production could drop from its present level of 560 mt (11 mbd) to 490 mt (9.8 mbd) by 2035.

Behind these numbers is a steady deterioration of the Russian well stock and reserves. The inventory of prospects is severely depleted: it contains "practically" no new large discoveries and an "extremely small" number of smaller cost-effective fields. The quality of existing reserves has also declined; a large part of the stock available for development consists of so-called TRIZ—hard-to-develop reserves (*trudnoizvlekaemye zapasy* is the official phrase)—with high development costs and low yields. The share of TRIZ fields is rising rapidly, but these fields tend to get less attention from the large companies, and as a result output from TRIZ has risen by only 6 mt in the last decade, reaching 38 mt in 2018, or less than 7 percent of total output.

New oil (greenfields), together with some increase from the TRIZ, has made only a modest contribution to growth, about 83 mt per year over the last decade. When this number is compared to the net increase in total annual Russian production of 61.7 mt over the same period, it suggests that output from the mature field stock (brownfields) has declined by 21 mt per year in ten years. In short, at this moment the oil industry is still staying ahead of the decline rate, but only barely.

For the present, low operating costs are keeping Russian conventional oil competitive at today's oil prices. Indeed, according to deputy energy minister Pavel Sorokin, Russian oil today would be profitable even at prices of $25 per barrel. But he warns that the costs of Russian oil are rising, especially those of new oil. The so-called TRIZ will require a world oil price of $70 to $75 to break even.[32] This cost trend, when combined with the prospect of lower oil prices in a world of declining oil demand, will put the Russian oil industry under a steadily tighter profit squeeze.

What needs to be done? Significantly, on this question the *Energy Strategy* is virtually silent. According to a thoughtful critique by two leading Russian analysts, Tatiana Mitrova and Vitaly Yermakov, the chief reason is the "difficult politics" surrounding the document, caused by a lack of consensus over the road ahead. In the meantime, oil policy over the last five years has consisted of "short-term reactive adaptation."[33] These are basically of three sorts: technological, fiscal, and diplomatic.

Technological Responses by the Industry:
Fracking and Horizontal Drilling

Faced with these threats, the oil industry has not been standing idle. Three major technological innovations over the last twenty years have gone a long way toward keeping production growing, mainly from the brownfields: the growth of horizontal drilling and fracking, combined with a technique called seismic visualization, which enables operators to see where they are drilling and where oil is likely to be located.

Horizontal drilling is a recent arrival on the Russian scene. Even though the first horizontal well was drilled in 2000, for the next twenty years the oil wealth of West Siberia required little more than conventional drilling. That has now changed. In the last decade, horizontal drilling has rapidly become a standard technology in Russian field operations. In 2018 nearly half (48 percent) of all development drilling consisted of horizontal wells. Rosneft has been a leader in the growth of horizontal drilling, having tripled its activity since 2013. Much of this activity is concentrated in West Siberia.[34]

Fracking too is a relatively recent arrival in Russia. Fracking was initially controversial in the Russian oil establishment, and as late as 2010 fracking was not yet in wide use. But in the next decade fracking took off. In 2018 the Russian oil companies spent 133 billion rubles (about $2 billion) on fracking, more than four times the 2010 figure.[35]

The widespread adoption of horizontal drilling and fracking in the last decade speaks well for the ability of the Russian oil industry to adopt new technology, but it does not change the basic picture of decline in the Russian brownfields. An eloquent example is the recent campaign by Igor Sechin and Rosneft to reverse the decline of production at West Siberia's fabled Samotlor field, once the workhorse of the Soviet oil industry, but reduced by 2015 to a paltry 15.5 mt. By 2019, Rosneft had succeeded in increasing Samotlor's annual output to nearly 19.4 mt—but at the price of increasing drilling nearly tenfold.[36]

Moreover, the industry remains highly dependent on foreign suppliers and operators for these new technologies. According to a 2019 report by the Ministry of Industry and Trade, 70 percent of horizontal drilling (presumably including most of the recent long-reach wells) relies on foreign suppliers; in fracking the dependence is 85 percent and above, especially for the latest technique, multistage fracking.[37] Thus, one of the major uncertainties affecting the outlook for Russian oil production will be the rate at which the industry reduces its dependence on foreign suppliers and modernizes its service industry. The Russian companies and policymakers gloomily accept that Western sanctions may well remain in force indefinitely, and indeed are likely to be stiffened. While at present they do not apply to providers of services for conventional oil, such as horizontal drilling and fracking, they may well do so in the future. The Russian companies are trying to cut their dependence on foreign expertise, notably by acquiring the Russian operations of established Western service providers. But this is only a palliative. The longer-term fortunes of the industry ultimately depend on the response of the Russian manufacturing sector as a whole to the challenge of producing up-to-date oil service equipment. The picture so far is not encouraging.[38]

Fiscal Responses: Oil Companies vs. the Ministry of Finance

As the costs of new oil rise, the Russian oil companies are increasingly pressing the government for tax relief and subsidies. More than half of Russian oil production benefited from some sort of tax preferences in 2018, compared to only 28 percent in 2014. According to the Ministry of Finance, these cost the Russian budget 1 trillion rubles (or about $16 billion) in lost revenues in 2018.[39]

The main opponent of fiscal relief has been the Ministry of Finance. The influence of the Ministry of Finance over the past twenty years has been one of the major stories of the Putin era. Shortly after he became president in 2000, Putin named Aleksey Kudrin, a close colleague from his days in the Saint Petersburg mayoralty, as minister of finance. Over more than a decade, with Putin's strong support, Kudrin restored

Russia's finances, which had disintegrated under Gorbachev and Yeltsin. He paid off Russia's external debts, balanced the budget, reformed the tax system, and created a rainy-day fund to store Russia's oil revenues. His resolutely conservative policy has been faithfully continued under his successor as minister, Anton Siluanov, and under Putin's new prime minister, Mikhail Mishustin, a tax expert who also comes from the Ministry of Finance. With one ear, Putin listens to the oil industry; with the other he listens to the Ministry of Finance.[40]

Faced with mounting pressure from the oil industry for more tax preferences for new oilfields, Putin in the fall of 2019 declared a moratorium on all new tax breaks and ordered an inventory of Russia's undeveloped fields. More than 700 fields, comprising over 90 percent of Russia's known proven and probable reserves, were reviewed. The results confirmed the alarm signals from the Ministry of Energy: one-third of the inventory, most of which was located in West Siberia, would be unprofitable to develop without further tax exemptions and subsidies.[41] Without them, Russian oil production could fall by as much as 40 percent, to 340 mt per year. Oil investment would fall to 2.3 trillion rubles, while the state's revenue from oil would decline to 4.1 trillion rubles.[42] To maintain production, Russia needed to double down on West Siberia, but that too would require more tax concessions. The Ministry of Finance was adamant—no more tax breaks. Even Sechin was told he could have only a fraction of what he was asking for. (Indeed, since then, as a result of the pandemic, Rosneft has had to pay additional taxes to the Treasury to help balance the government's budget.)

The industry has come to expect these preferences (as the Russians call them), and consequently every major new investment project is preceded by intense negotiations. The Ministry of Finance attempts to hold the line,[43] while the companies multiply their appeals, often to the president himself. The result is a constant dance, which only intensifies when oil prices go down. In the wake of the COVID pandemic, the drop in state revenues has been even sharper.[44]

Diplomatic Responses: Cooperation with OPEC

In 2016 Russia and several non-OPEC producers pledged together with OPEC to cut oil production by a combined 1.8 mbd starting in January 2017. The aim is to keep world oil prices high enough to enable the main oil producers—chiefly Russia and Saudi Arabia—to maintain what is known as the fiscal break-even price—that is, the price at which they can balance their budgets while maintaining their social welfare systems and state investments. Here Russia would appear to have an advantage: Saudi Arabia reportedly requires a world price in the $70s to balance its budget, while Russia needs only a price in the $40s.

But the arrangement has proven unstable, as became clear in early 2020 when trust between the two main partners broke down and a price war ensued. Oil prices collapsed again, forcing Russia and Saudi Arabia to hastily restore their agreement, with even larger production cuts than before. The April 2020 OPEC+ "mega deal" committed both Russia and Saudi Arabia along with other members of the Vienna Alliance to a combined production cut several times larger than any reduction previously attempted.[45] But the episode demonstrated how unstable such agreements are likely to be. The accord has been unpopular with the Russian oil companies, especially (as noted above) with Igor Sechin, the CEO of Rosneft, who has criticized it publicly and repeatedly.[46]

For Putin, the main objective was to protect Russia's oil revenues, and especially the state's share, while helping to rebalance the global oil market following the record decline of demand and oil prices caused by the COVID pandemic. (As we shall see below, when oil prices go up, the state's share increases; but when oil prices drop, it is the companies' share that grows.) Sechin's main goal was clearly to discourage North American tight-oil producers and to limit the freedom of action of the United States in global oil markets.[47] As a side benefit, lower oil prices would maximize the Russian companies' share of revenues versus that of the state, thus helping to provide badly needed cash flow for Rosneft's investment program, particularly Sechin's ambitious plans for the Arctic, described below. Note that the two men did not disagree over climate change or "peak oil demand"—neither one believed in it—but rather over the best way to monetize Russia's oil in the near term.

The bumpy history of the OPEC+ accords offers a preview of what may happen ahead, as the world's largest oil producers are pitted against one another in an increasingly desperate zero-sum competition for the world's dwindling oil demand. All sides, and especially the weaker members of OPEC, will have a growing incentive to break away from any agreements to curtail production.

Summing Up the Short-Term Responses

The key point about these three responses is that they are all short-term expedients ("short term" meaning about a decade). As output from West Siberia declines, the average lifting cost of Russian oil will rise. Lower well productivity will require increased inputs, such as electricity for downhole pumps, as already mentioned. As the industry moves out of West Siberia, transportation costs will also rise, as the industry works in more remote locations, requiring more air transportation, as well as ground transportation over roads that do not yet exist, or ice roads that are melting. The increasing share of new fields in new provinces will require more investment; and because such fields tend to be less productive, this will cut profit margins and add to pressure for more tax breaks. There will be growing limits to what the government can grant to the oil industry; according to the Ministry of Finance, by 2033 more than 90 percent of Russian oil investment will require special tax breaks from the government.[48] The substitution of domestic inputs for imported ones will continue, but the easy opportunities have been exploited and further progress will be slow. Finally, the OPEC+ agreement with Saudi Arabia may not survive long. In short, the expedients of the last decade will have a decreasing impact ahead.

The oil industry has been slow to rethink its basic strategy, which has been essentially to continue focusing on conventional production from West Siberia while appealing to the government for help with new projects. After the euphoria of 2000–2014, when production and prices seemed to rise without end, Russian companies and policy-makers are finally beginning to focus on the darker outlook, but they have no agreement on the way forward.

There are broadly only two longer-range ways to go—deeper into zones of unconventional oil, or outward into the Arctic offshore. The Russian companies are beginning to investigate both avenues, but each has been surrounded by controversy, and the early results have not been encouraging.

Potential for Surprise?

The main potential surprise lies deep in the ground, in the form of "tight oil," also known as shale oil. What are the odds that Russia could experience the same explosion of unconventional tight oil as the United States? Several Russian companies, notably Gazprom Neft, are actively investigating unconventional tight-oil prospects. The biggest potential prize is the Bazhenov, which underlies much of the West Siberian oil province. The Bazhenov is estimated to contain up to 74.6 billion barrels of technically recoverable tight oil.[49] However, investigation of the tight-oil potential of the Bazhenov has been slowed by the imposition of US sanctions in 2014, which caused Exxon to withdraw from a joint venture to drill in the Bazhenov. So far there has been little progress to report.

The phenomenon of tight oil requires a word of explanation. The basic idea behind tight oil is that oil evolves. Over millions of years, it is slowly transformed from decaying organic matter into an interme-diate substance called kerogen, which in turn serves as the source for oil and gas. At the high temperatures prevailing underground, the oil migrates upward and ultimately turns into gas, which escapes. There is thus a brief "oil window," during which oil may be present in the source rock. The United States turns out to be lucky: formations such as the Permian Basin in Texas are right in the middle of their historical oil window. So is the Middle East. Where Russia may be on the clock—that is, how mature the Bazhenov source rocks may be—is as yet unknown, although early indications from Russian experts suggest that the Russian potential may be more limited than that of the United States.[50] There are signs that the central core of the Bazhenov, the area with the richest source rock, may be in the gas window, which follows the oil window. If that turns out to the case, then the Russians may be a few million years too late to get much oil out of the Bazhenov.[51]

Whatever the Russian geological story turns out to be, US experience suggests that aboveground factors are just as important as geology in determining the course of tight-oil development. Small, highly entrepreneurial independents played the key roles in the US story, attacking each prospect by trial and error until they cracked the code for each one. Every success, in other words, is custom-made, a formula to which large companies are poorly suited. In addition, the independents benefited from a favorable environment—a flexible and innovative service sector, a convenient transportation system, supportive regulation, and above all, cheap and abundant finance. None of these features are present in Russia.

As the Russians watch the US experience in the Permian Basin, they must find much to dislike about it. For example, fast decline rates (the Russians are used to long lifetimes and long tails); high requirements for advanced fracking and horizontal drilling equipment (which at this moment are mostly still provided by foreign companies); a constant requirement for finance (credit is very expensive in Russia, and foreign capital is largely unavailable); custom design of each well, to find the sweet spot in each case (the Russians prefer "one size fits all"); the need for small, risk-taking companies (the Russians don't have such small companies, and the large companies don't like risks, unless the government pays for them); and above all, the fact that tight oil is not cheap oil, and the US tight-oil business has yet to show a profit. For all these reasons, tight oil has an uncomfortable position within the traditional structure and expectations of the Russian industry. In sum, the prospects for a tight-oil surprise in Russia are dubious.

Russia's Missing Bluefield Oil and the Battle for Putin's Ear

The other potential surprise in future Russian oil production could be oil from Russia's Arctic offshore, which runs for thousands of miles along Russia's northern coast. This has recently been the subject of lively debate within the government, prompted by the prospect that a peak in global oil demand, followed by lower demand and oil prices, could leave the Arctic offshore stranded indefinitely.

Leading the charge for an aggressive offshore program was deputy prime minister Iury Trutnev, Putin's representative for the Russian North and Far East. He accused Gazprom and Rosneft, which hold a monopoly of Arctic oil and gas exploration licenses, of systematically underinvesting in exploring the Arctic offshore. Yet unless Russia began developing its Arctic offshore resources soon, Trutnev argued, it would find itself sitting on a stranded asset. Said Trutnev at a conference in 2019: "I think it's just strange to wait for what happens next. Oil isn't a resource whose price won't change. . . . Humanity is continuing to work to reduce consumption of energy resources and motor fuel. . . . The longer we don't use it, the more it'll cheapen."[52] Challenging Gazprom and Rosneft, Trutnev proposed creating a new company, which he described as "a Norwegian model," into which foreign investors could be invited.

Trutnev's proposal was quickly rejected as unrealistically ambitious, but Trutnev had performed a useful service by training a spotlight on the key dilemma for Russian oil strategy. If West Siberia is fated to decline, and if tight oil does not come to the rescue, then where will the next generation of Russian oil come from? Igor Sechin, the ever-formidable CEO of Rosneft, was ready with an answer: stick to the shallow Arctic near-offshore, and build out from already-established infrastructure there. In Sechin's vision, the key is to move gradually eastward along Russia's northern coastline, building infrastructure along the way, mainly onshore. But Sechin's plan comes with a high price, funds for which can only come from massive government investment and subsidies. In a meeting with Putin in April 2019, Sechin made his case. His so-called Arctic Cluster would cost 5 trillion rubles (currently about $68 billion) in state support over ten years, plus a special tax package for thirty years, in exchange for which Sechin promised the president an additional 100 mt / year in new oil production.

For Putin, Sechin's proposal must have sounded appealing from several vantage points, not least of which was Sechin's promise to evacuate the new oil toward Asia via the Northern Sea Route (see Chapter 8 on the Arctic). Yet the economics of the Arctic Cluster are highly uncertain. Its reserves have not yet been confirmed. The infrastructure is at present almost entirely lacking. The cluster will require a new port, which will need to be connected to both new and existing fields by a network of pipelines. But who will pay for these? Rosneft warned that without fuller state support the cluster would be unprofitable to develop. Yet the Ministry of Finance is resolutely opposed, and Putin, as we have seen, has hesitated.[53]

The Arctic offshore is where the high costs of new oil collide with the prospect of lower global prices. Opponents of Arctic oil development argue that it cannot be profitable at less than $100 per barrel. The net result is an increasingly dim outlook for Russia's bluefield prospects. Even before the onset of the pandemic, the official Russian targets for production on the Arctic shelf had been scaled back. In the latest available official scenario, output is projected to reach only 9–11 mt per year by 2035.[54] This is barely more than Russia is producing there today.

In sum, the two main avenues of long-term progress—tight oil and Arctic offshore blue oil—are problematic. To be sure, surprises may await. Or Igor Sechin's solution, to focus on the shallow Arctic coastline, could represent a temporizing way forward. But lacking some kind of breakthrough, the Russian oil industry will be forced to tread water, or more likely, to retreat step by step as West Siberia declines.

Who Will Take the Hit?

The implication of these trends is that the costs of producing Russian oil—which hitherto have remained remarkably low, thanks to the continuing contribution of the brownfields—will inevitably rise in coming years, as the brownfields decline and the share of greenfields slowly increases. Even so, Russian oil will still be relatively low-cost compared to most of its competitors, and Russia will always find a buyer for its barrels, even in a world of declining demand. But with Russian oil occupying a higher place on the oil-cost curve, its profit margins will be squeezed. The question is: What will be the impact on Russian export revenues, and on the division of oil income between the industry and the state?

In a world of declining oil demand and prices, the state would take most of the hit. The Russian oil tax system is based on a sliding scale, keyed to world oil prices. The basic concept is straightforward: as oil prices go up, the state's share increases; but as oil prices go down, it declines.[55] Thus in 2011, when world oil prices briefly exceeded $120 per barrel, the state's tax take was close to 80 percent; but by 2016, when oil prices fell below $40 per barrel, the government's take declined to about 40 percent.[56] The intent is to ensure that the oil industry will have the resources to cover operating expenses and capital investment, independently of price movements. The state, in other words, takes most of the oil margin when prices are high, but takes the hit when they drop.[57] Over the coming decades, as Russian production declines and export prices drop, the state will be the biggest loser.

What might be the net effect on Russia's oil income of all the points discussed so far in this chapter? As a very rough thought experiment, let us assume that Russian oil production by 2050 declines to 450 mt, or 9 mbd. Assuming further that Russia continues to export about 70 percent of its oil output, at a world price of $30 per barrel, this would yield about $75 billion, about 40 percent of what Russia earns today. Of that amount, if the share of the state declines to one-third, that would leave the state with about $25 billion, or about one-third of its present take.

The state will face a growing dilemma—whether to continue the policy of supporting the oil industry with tax preferences and the sliding scale described above, or to capture as much of the oil rent as possible quickly—essentially to "take the money and run." This dilemma will only grow more acute as the 2030s begin.

So far we have focused on the outlook for the upstream, but the downstream deserves mention as well, particularly for its impact on domestic consumption and automobile ownership. As electric vehicles spread around the world, demand for Russian refined products will decline, especially gasoline and diesel. Gasoline exports, though traditionally modest, had been expected to grow strongly; the Russian Energy Strategy to 2035 projected a quadrupling; diesel exports, while traditionally much larger, had been expected to double. These will not happen. Instead, excess gasoline and diesel will be rerouted into the domestic market, causing prices at the pump to decline.

This brings us to the subject of electric vehicles (EVs). It takes a brave auto-owner indeed to drive an EV in Russia. The infrastructure is almost entirely lacking: In all of Russia there is a grand total of 122 charging stations, of which 71, not surprisingly, are clustered in Moscow and Saint Petersburg. In the entire Far East, there are only 2. Two sorts of Russians own electric vehicles: half are well-heeled customers in the two capital cities, who buy Teslas and other luxury brands as prestige curiosities. In the east, in contrast, people buy used Nissan Leafs imported from Japan. All told, there are barely 11,000 electric vehicles in the entire country, almost all passenger vehicles, compared to a nationwide total of 54 million cars with internal-combustion engines.

It is a safe bet that electric vehicles will not occupy more than a niche market in Russia in the foreseeable future. There are three obstacles:

roads, trains, and cheap fuel. There are hardly any intercity highways (although one exception is the Moscow–Saint Petersburg highway, which actually has ten charging stations). Without roads, there is little role for trucks; in Russia almost all freight goes by rail, and will undoubtedly continue to do so. As for gasoline, it will only become cheaper at Russian pumps as international demand declines. In short, EVs in Russia face an insuperable uphill battle. The Russian government, recognizing this, has made little effort to promote them. The internal combustion engine still has a long life ahead of it on Russian roads.

Conclusions

The Russian oil problem can be summed up quite simply: Russia is squeezed between the threat of lower production and higher costs at home, and the prospect of lower oil demand and lower prices abroad. The first has little to do with climate change; the second is increasingly driven by it as the developed world moves toward decarbonization. If there is a third narrative—"slower for longer"—as a result of the COVID pandemic, the momentum behind decarbonization could slow, but because of lower global GDP growth rates and diminished prosperity, this narrative is not necessarily good news for Russian oil either.

The chief uncertainties in the global oil market, as we have seen, lie outside Russia, and hence outside of Russian control, being driven primarily by trends in technology, economics, and demography, influenced to varying degrees by the energy policies and regulations of other states, particularly oil buyers. Over these, Russian diplomacy will have little leverage. The sole stopgap exception is oil-producer agreements such as the OPEC+ accords, which are likely to be unstable and may not last.

The other source of pressure on Russia's oil revenues comes from inside the country. Production from the West Siberian brownfields will continue to decline, despite the efforts of the oil companies. There will be continued movement to the greenfields, but these will be less productive than the traditional brownfields, and total costs of new greenfield oil (that is, capital costs plus operating costs) will be far higher,[58] while the Arctic offshore—the bluefields—will remain out of reach.

Russia's brownfields, with their low operating costs, will remain profitable, but by the mid-2030s their share in total production could be less than half, while the export break-even price of new oil from greenfields may well be above world prices.[59] Thus, when Russians say, as they frequently do, that Russian oil will remain profitable at $25 to $35 per barrel, that is only half true. The higher costs of the greenfields will be hidden by the many subsidies and concessions granted by the state, and even though this may keep the oil companies profitable on paper, this will come at the state's expense, and there will be no concealing the underlying verdict: the net value of oil exports to Russia will decline. This will be true even if the companies are able to maintain production at close to today's level of 11 mbd, because the additional production will inevitably be at the higher end of the cost spectrum.[60] As costs mount, the export break-even price—the price that oil exports need to fetch in order to break even—will inch steadily upward, and as it does, the profit from exports, to the companies and especially to the state, will shrink. This would happen even if global demand and prices remained at their present levels.

As the decline in global oil demand takes hold, however, several things will happen. First, demand is likely to go down faster than supply, as states that depend on oil revenues will be reluctant to cut production, and they will compete strongly with one another through their state-owned companies, putting additional downward pressure on prices. Second, demand will be strongly influenced by tipping points at the margin, such as the point when electric vehicles fueled by renewables finally become dominant, or when plastics become subject to worldwide bans. Oil demand at such points could begin to drop rapidly. Third, oil investment will decline in response as investors look for more attractive opportunities, although ongoing advances in oil-field technology and efficiency will tend to brake any sharp fall in supply. By mid-century, all of these trends will be well under way, resulting in an environment of low and declining oil prices in the 2050s and beyond, even in the slow-transition narrative.

The slow-transition narrative, described earlier, would seem at first glance to be the most favorable for Russia because it projects two more decades of growth in global oil demand and a slow decline after the peak. A slow-transition scenario would give the Russian oil industry more time to apply new technologies, perhaps including a breakthrough in tight-oil technology. Sanctions might be relaxed and international oil companies might return, finally enabling progress in the Arctic offshore. More time would give Russia more latitude to prepare for the final decline of oil.

The fast-transition scenario would clearly be highly unfavorable. Because the forces that drive it originate mostly outside Russia, it would have little control over them. The only near-term response available to preserve oil revenues would be to engage in an all-out battle for market share at lower prices; this in turn would be possible only so long as low-cost production from West Siberia can be maintained. In a fast-transition scenario, Russia's competitive position rapidly worsens.

The wild card is the long-term impact of COVID-19, which, as discussed earlier, could result in a third narrative—slower for longer—consisting of a mixed transition over a more extended period. But this is not a reassuring narrative for Russia either, if the result is lower global GDP growth, changing work habits, and an increasing focus on domestic energy sources. The likely outcome would be lower oil demand at lower prices over an extended period. The refinery industry would be particularly affected by changing consumer habits at the pump; an early signpost is the current severe overcapacity in the global refinery sector. If this pattern persists, it will cut into Russian product revenues even more than into export revenues from crude oil.

The chief issue, to repeat, is not supply, but demand. The Russian oil industry has colossal oil reserves, and Russian oil will long remain some of the lowest-cost oil in the world. But Russian revenues from oil exports will depend above all on the future of oil demand, together with the price of alternative sources of energy. Russian oil will not be "stranded" any time in the foreseeable future. The issue is how much money it will bring in.

But what alternatives are there to oil? That is the subject of the following chapters, beginning with the outlook for natural gas.

3

CAN NATURAL GAS REPLACE OIL?

Leonid Mikhelson, the founder of Russia's liquefied natural gas (LNG) company Novatek, is Russia's third-richest man, with a net worth estimated at over $23 billion in 2020.[1] He is also that rarity in Russia, the creator of a successful private-sector start-up. This has attracted the Kremlin's active support.[2] In a short time Mikhelson has become the face of Russian LNG, Putin's entry into the future of Russian gas.

Mikhelson started his career as a builder of gas pipelines in Kuybyshev oblast (now Samara). "My father was the head of the largest construction unit in the Gazprom system. He took me everywhere with him, wherever the company built pipelines. I used to go to gas and oil fields before I went to school."[3] By the late 1980s, Mikhelson had succeeded his father as the head of the Kuybyshev pipeline construction trust. He quickly realized the opportunities presented by the post-Soviet privatizations. In 1991 his company was one of the first Soviet-era enterprises to privatize, under the name Nova, subsequently changed to Novatek.[4]

There was no money to be made from pipeline building in those years, so in the mid-1990s Mikhelson moved to West Siberia,[5] where he began acquiring undeveloped gas fields, which at the time could be had virtually for the asking.[6] During the 2000s, Novatek made its mark by selling gas to industrial customers and by exporting condensate—the light oil that comes up with natural gas—to refineries in Europe, where it fetched a good price, using new technologies to reach deep liquid-rich formations. Mikhelson, a born entrepreneur, began making plans to take the next step—to develop LNG. This attracted the interest of the French energy company Total. In 2011, with then-premier Vladimir Putin looking on, Total signed an agreement to buy a 12 percent interest in Novatek and a 20 percent stake in Novatek's newly formed subsidiary Yamal LNG, based on the Yamal Peninsula.[7]

Russia being Russia, Novatek could not have reached its position without strong state support and key sponsors. From the beginning Mikhelson proved adept in forming political alliances and defusing potential opposition. He benefited from the fact that at that point, Gazprom, then still Russia's gas monopoly, remained largely indifferent to LNG and continued to focus on its traditional business of pipeline gas. Vladimir Putin, who early in his presidency had sought to make Gazprom Russia's main energy champion,[8] subsequently changed his view and began backing Mikhelson and Novatek as his chief vehicle for LNG.[9]

The vision of Mikhelson and his French partner was that LNG from Yamal could be exported both east and west—west to Europe during the winter, and east to Asia during the summer via the Northern Sea Route through the Arctic Ocean, with the assistance of nuclear-powered icebreakers. This was a daring concept, because the eastward route through the Arctic had never been used for LNG.[10] Nevertheless, from 2014 on, thanks to the partnership between Novatek and Total, and the strong support of the Russian government, progress was rapid. In 2017 the first LNG tanker, named after the late CEO of Total, Christophe de Margerie, who had tragically died in a freak airplane accident in Moscow, was dispatched to market in Asia.[11]

That maiden voyage marked a turning point in Russian gas strategy. For the past half-century Russia's strategy has rested on pipeline exports to Europe. But the traditional model is now under threat from Europe's climate-change policies. Meanwhile, gas demand in Asia is booming. Russia is now attempting to shift its gas exports to Asia, using LNG as its spearhead, together with a new pipeline to China. Although

President Putin rejects the peak oil narrative and insists that Russia's oil revenues will remain strong indefinitely, he praises gas as Russia's most promising export fuel of the future.[12]

In short, Russian gas policy is at a crossroads. But will this strategy succeed? Can the growth of Russian gas revenues from Asia make up for the coming decline from Europe? And if Russia's oil revenues shrink in the future, can profits from gas help to make up the difference?

The answer to these questions is no. My argument rests on three points. First, though Russia has abundant supplies of gas, its problem is not on the supply side but on the side of global demand. In particular—and this is the second point—Europe's increasingly stringent decarbonization policies will soon drive Russia's traditional European market into decline. Third, though gas demand in Asia will grow strongly, Russia will face growing competition from rival suppliers, and ultimately from renewables. As a result, revenues from its exports to Asia will not be enough to offset the decline in Europe. The bottom line is that despite Russia's enormous wealth in gas, its export revenues from gas will grow only slowly, if at all, and they will not be able to offset the decline in income from oil.

The next three sections explore these points in more detail.

Russia's Abundant Supplies

Gas was the last major success of the Soviet period. Post-Soviet Russia was left with a priceless legacy: the world's then-largest gas industry, practically brand-new, backed by the world's largest gas reserves.[13] Half of the economy had already been converted to gas. Gas, even more than oil, saw Russia through the 1990s.[14]

But by the mid-2000s the Soviet-era gas legacy was running down. Russian gas was still based on the resources and infrastructure inherited from the Soviet era. Most of its production came from just three fields in West Siberia, which by then were in decline. But in 2006, under Vladimir Putin, Russia launched a new gas offensive, based on the Yamal Peninsula in the north of West Siberia. Simultaneously, it expanded construction of a new series of gas export pipelines, designed to bypass Ukraine and provide more diversified access to the European market.[15] At the same time, as part of its "pivot to the East," Russia has developed two new giant fields in East Siberia and has built a first gas pipeline to China, eventually to be followed by others.[16] In parallel, as noted above, Russia began developing LNG, also from the Yamal Peninsula, adding to existing production from Sakhalin Island in the east.

In short, in the space of fifteen years Russia under Putin has developed a new generation of gas.[17] Consequently, Russia will remain abundantly provisioned to mid-century and beyond. Climate change will have no discernible effect on supply, although melting permafrost will increase costs. The main uncertainty is how much of this prospective capacity will actually be used. We will turn in a moment to the export market, but first we need to take a look at domestic consumption. How much of Russia's gas will be consumed domestically, and how much will be available for export?

Russian gas, unlike oil, is largely consumed at home. In 2019, 70 percent of total production was used domestically, while only 30 percent was exported.[18] This is the result of an explicit policy, which dates back to Soviet times, to support the population and the economy with cheap gas. Today most domestic gas is still sold at prices well below global levels. It is a core part of the rent that the Kremlin passes on to the population as part of an implicit social contract to secure its support, and that is likely to remain true for the foreseeable future.

This has consequences. Gas is overconsumed throughout the economy, particularly by the residential sector. Industry, too, benefits from low-priced gas, and has little incentive to modernize inefficient gas-consuming production processes. It is hardly an exaggeration to say that the entire political and economic system is built around cheap gas.[19]

What about the prospects for greater efficiency? From 1999 to 2007, when Russia enjoyed high oil revenues and Russian industry made good profits, it invested part of them in improving the efficiency of gas consumption. In contrast, the years since then have been ones of relative stagnation. Russian industry today lacks the capital to renew its stock and further increase its efficiency. There has also not been much improvement in the efficiency of residential consumption. The one exception is the power sector, where new gas-fired power plants are gradually replacing Soviet-era stock. In short, for the foreseeable future Russia will remain a predominantly gas-fired economy that overconsumes gas.[20]

Is there a potential for a significant change in gas's share in the economy, such as would occur if there were a large-scale transition to renewables? As we shall see in Chapter 5, on renewables, the low domestic price of gas gives it a commanding advantage, and any significant displacement of gas by renewables is highly unlikely.

Finally, how might climate-change policy affect domestic gas consumption? Because of the large role that gas plays in Russia's primary energy demand, it is the main source of Russia's CO_2 emissions. The power and heat sectors are essentially gas-fired, as is industry. The only major exception is the transportation sector, which burns gasoline and diesel, and the power sector in the east, which is fueled with coal. Much of the opportunity for Russia to cut back on greenhouse emissions, therefore, lies in the domestic gas sector. But because of the economic and political inertia of the present consumption pattern, this situation is unlikely to change.

In sum, inside Russia there will be only slow growth in domestic gas demand, in contrast to Russia's abundant supply. In other words, much of Russia's new gas supply will be available for export. But will there be demand for it? What is the outlook for global gas, and how will it be affected by climate change and the energy transition?

The Outlook for Global Gas

The next thirty years, practically all experts agree, will be a time of record strong growth for natural gas demand. Gas, which at present accounts for about one-quarter of global energy demand, will overtake coal in the 2020s and oil by the 2040s. On this view, by mid-century gas and oil will be tied as the two leading energy sources in the world. Oil will be declining, but gas will still be growing. In absolute terms, global annual demand for natural gas by 2050 could nearly double. As the century moves on, gas's dominance over oil will continue to increase. Gas's only competitor will be renewables.[21]

This at least was the picture on the eve of COVID-19. The pandemic has introduced new uncertainties. It could cut the growth of both long-term gas demand and long-term supply. On the demand side, the yearly increase in global gas demand to mid-century could be less than 1.0 percent, instead of the vigorous 2.5 percent annual growth rate of the last two decades. In Asia, coal could survive or even prosper as a serious competitor to gas into the mid-2030s. In Europe, gas consumption could be held back by slower economic growth and continued reliance on coal in some countries. The latest forecasts of global gas demand have grown noticeably more cautious. BP, for example, projected as recently as 2019 that gas demand would continue to grow strongly into the 2040s, but it now believes that gas demand could peak by 2030, and by 2050 it could fall back to today's level and decline rapidly from there.[22]

On the supply side, the chief uncertainty concerns the growth of LNG. Pre-pandemic scenarios projected a strong continued expansion of investment in LNG; a large number of LNG projects were nearing final investment decisions, which could have totaled as much as $250 billion of investment by the mid-2020s. LNG production capacity was expected to nearly double by the 2040s, from around 365 mta in 2019 to over 700 mta.[23]

The impact of LNG was compounded by the development of shale gas in the United States. The resulting "shale gale" abruptly transformed the United States from a gas importer to an exporter, mainly via LNG. Together with other LNG exporters, such as Qatar and Australia, the United States took aim at the traditional dominance of pipeline gas in Europe, threatening established exporters, especially Russia.

Such was the boom in LNG that it became widely expected that the 2020s would be an era of LNG oversupply and that the result would be downward pressure on prices, not only of LNG but of gas generally.[24] It was widely predicted that the next thirty years might be the golden age of gas, but they would not be the golden age of gas profits.

But the pandemic caused a wave of cancellations and postponements. One lingering consequence of the pandemic could be a reluctance to commit capital to large projects such as LNG.[25] As a result, global LNG capacity in the 2030s and 2040s could be considerably lower than projected earlier. Yet as the world recovers from the pandemic, continued technological advances and efficiency gains in LNG production and capacity will once again drive increases in supply at lower costs. The supply of LNG is likely to resume growing, if perhaps not as strongly as before.[26] But what will happen to demand?

Climate change will play a role in two ways. Much of the developing world views climate change as less of a problem than pollution. Natural gas is still regarded as a "virtuous fuel" because it is cleaner than coal; therefore, concerns over air quality will drive continued increases in gas demand, particularly in Asia. In Europe, on the other hand, climate change combines with classic environmental concerns to drive energy policy toward decarbonization. Gas has lost the aura of a clean fuel, and the search is on for alternatives.

Under all scenarios, gas faces two major battles ahead. The first, which will dominate the 2020s, is with coal. The chief battleground will be China and India, which together account for 60 percent of global coal demand. The contest will be especially stiff in the power sector, and the outcome will depend on a complex mixture of economics and politics. On the one hand, new coal-fired power plants are still being built all over Asia. Many have high operating costs, however, and at present are frequently idle. But much of this coal-fired fleet is still very young; consequently there is considerable political resistance to curtailing it. Major political and social interests are at stake, from Australia to China, which will tend to keep coal supply and demand strong.[27]

In the developing world, gas is typically disadvantaged by its higher delivered cost compared to coal, as well as the lack of distribution infrastructure; consequently, a strong policy push is required to displace coal.[28] Political elites are ambivalent: all agree on the harmful effects of pollution from coal, but they worry about the social and political consequences of shutting down coal production. In China, in particular, policy has swung unpredictably back and forth from one concern to the other, as I discuss below.[29] Yet despite these mid-range uncertainties, the path for coal heads inexorably downward.

Beyond 2030, the one remaining rival to natural gas will be renewables. Together, renewables and gas will account for most of the growth in primary energy demand. The main uncertainty is the relative share of each. In Europe and the United States, renewables will gradually overtake gas, although they will be slower to penetrate the heat and cooling sectors than the power sector (this is discussed below). The chief remaining battleground will be Asia, where renewables are already rapidly gaining ground, being greener and, increasingly, cheaper than gas.

How do these global trends affect the prospects for Russian gas exports? In a world of abundant supply and sharp competition among suppliers, the Russians have the advantage of low lifting costs, but that will not be enough. Take the pipeline trade first. Here Russia has major advantages because of its legacy transportation system, augmented by a vast new network of export pipelines. Thanks to its investments in new West Siberian gas and export pipelines, Russia is the lowest-cost shipper to Europe, compared to LNG, and this will likely remain the case for the foreseeable future.[30] However, Russia will not enjoy the same advantages in its pipeline exports to China. Both its East Siberian fields and its new pipeline, Power of Siberia, have turned out to be very costly, and any successors are likely to be costly as well. The profits from the pipeline exports to China have been disappointing so far.

Transportation costs will likewise be the main problem for Russia's LNG. The largest part of the cost of LNG supply comes from processing and transportation. Russia's main resource for LNG, the Yamal Peninsula of West Siberia, is about equidistant from the European and Asian markets; it can thus swing both ways. The major obstacle, however, is on the eastern side, as Russian tankers must thread their way across the icebound Arctic Ocean, through the Northern Sea Route to reach Asia. As the ice continues to melt, Russian LNG to northern and central China should be competitive, but more distant destinations may be a stretch.

Thus, the two critically important gas markets for Russia are Europe and China. We now take a closer look at both.

Focus on Europe

The European Union's goal is to decarbonize, by making the EU carbon-neutral by 2050.[31] This implies a far smaller share for gas in Europe's primary energy mix, and lower gas demand. But it will easily take a decade before alternatives to gas are ready. Thus, there will be two phases to the evolution of the European gas market: during the first, through the mid-2030's, Europe will continue to need imported gas, particularly to support its battle to phase out coal. During the second phase, however, Europe's need for imported gas will decline, as gas loses out to renewables.

Let us look more closely at these two phases, to see what they imply for Russian gas strategy. The global abundance of gas at low prices, combined with higher carbon prices in Europe, will favor gas over coal throughout the 2020s, especially in the power sector. Political initiatives, such as the Powering Past Coal Alliance, will add to economic forces to produce a steady displacement of coal by gas, in what the International Energy Agency (IEA) describes as a "quick win" for gas and the final demise of coal in power. By the end of the 2030s the battle against coal will have been won. Coal-fired capacity will have declined from about 180 gigawatts today to less than 50 gigawatts in 2040. The elimination of coal may be slowed by low coal prices and rearguard political resistance by coal-based interests, but in the long run the disappearance of coal from most of Europe is inevitable.[32]

As coal is defeated, gas will take its place as the chief target of environmental opposition, and after 2030 the battleground will shift to a contest between gas and renewables. Here, too, there are two competing narratives—fast-track versus slow-track. The fast-track narrative is increasingly compelling, especially when factoring in the determination expressed by the EU to make the economy carbon-neutral by 2050. As the costs of solar and wind continue to plummet, the cost of building new capacity based on renewables will soon drop below the operating costs of gas-fired plants. Gas in the power sector will be increasingly limited to a "peak coverage" function—generating power on a standby basis when sun and wind are lacking. But that function too will fade, as progress in battery technology and more sophisticated demand-side management make storage of renewable power more economic than even the most efficient gas-fired power plants. By mid-century, demand for gas in the power sector will fade.

Outside the power sector, demand for gas will be slower to decline. The heating market accounts for about half of current gas consumption, and at present there is no ready alternative to gas to cover winter loads, although these may decline as the climate warms.[33] The main opportunity ahead is electrification of home heating, thanks to the widespread adoption of heat pumps, powered by renewable electricity and biomass. But the slow rate of replacement of existing housing stock will limit the rate of penetration of electricity-based heating, and gas will remain in demand. Gas will likewise be needed for industry, in sectors such as steel and cement. This is leading to vigorous efforts to find alternative fuels that could be blended with natural gas in the existing gas infrastructure, particularly hydrogen. But it will be some time before hydrogen can play that role, as discussed below.

On balance, Europe will continue to need imported gas, and Russia will be well positioned to supply it. But after a relatively favorable decade in the 2020s, gas importers into Europe will have to fight it out for market share in a gas market that by mid-century will be shrinking.

There is irony in the fact that Russia, having expended vast amounts of capital to develop a new generation of gas fields and pipelines primarily intended for the European market, now faces the prospect of a declining market in Europe. Russian gas strategists face a radical adjustment, away from Europe and toward China and the rest of Asia. But can Asia replace the European gas market, in volumes and in profits?

Focus on Asia

The gas story in Asia will be quite different from the one in Europe. While Europe remains—for now—the centerpiece of Russia's gas exports, the Chinese gas market is increasingly the focus of its export policy, and these two facts are likely to govern Russia's gas exports to mid-century. By the mid-2000s, gas was becoming a major priority of the Chinese government. Since then China's gas sector has grown rapidly. In 2018, consumption reached 280 billion cubic meters (bcm), mostly devoted to industry and residential heating; power accounted for only 17 percent. But despite its rapid growth, gas in 2018 made up only 8 percent of China's primary energy demand.[34] Increasing that share is one of the Chinese leadership's main energy and environmental objectives.

But demand has raced ahead of domestic production, which was held back for decades by a policy of paying low prices to producers, which made upstream investment unattractive. Production was further held back by the lack of an adequate transmission network. By 2018, Chinese gas production was only 160 bcm (of which only 11 bcm were of shale gas).[35] Consequently, in 2006 China began turning to imports; its first LNG terminal was built in the southern Guangdong Province, close to emerging gas demand. This was then followed by a major pipeline project, the West-to-East line, which brought gas from Turkmenistan to the populous eastern coast.[36] In 2018 China imported more than 90 million tons (mt), one-third of its consumption. Of this quantity, 60 percent was pipeline gas but 40 percent was LNG, and this share was growing fast.[37]

Chinese gas policy over the years has been inconsistent and unpredictable, and recent events are no exception. In 2017–2018, China adopted an ambitious policy to displace coal with gas; but in late 2019, alarmed by the slowdown in the economy and under pressure from the provinces, the government relaxed its coal-to-gas switching policy, and began touting the use of "clean coal" instead of natural gas for heating.[38] Pricing and transmission policies are a tangle of inconsistencies. For example, LNG importers (which are the three large national oil companies) lose money on imports, but they make it back through their control of transmission tariffs; as a result, at the end of the chain, retail gas prices are still too high for gas to displace coal. Consequently the gas sector is caught in a squeeze: importers and producers are unable to make money on gas, yet retail gas is too expensive to compete on the domestic market. Until China fixes its pricing and tariff policies, the penetration of gas into the Chinese economy will remain constrained, particularly in the northeast of the country, where the pipeline network is still undeveloped.

Despite these problems, China's leadership remains committed to increasing the long-term share of gas in the economy, especially in the power sector. The pricing system is under active review. Other reforms concern the infrastructure, especially for transmission. So far they cover only a portion of China's pipelines and receiving terminals, but they will create momentum for third-party access and encourage new pipeline construction, which should ultimately bring down prices in the domestic gas market. In December 2019 China finally created an independent national pipeline company, China Oil and Gas Piping Network Company, which is responsible for the operation of key oil and gas midstream infrastructure and new construction.[39] Reform is under way, but its impact will be only gradual.

It was in this setting that the Russians and the Chinese first began negotiating an import pipeline from East Siberia to northeastern China. On the face of it, this project fit squarely within China's priorities. Yet it took three decades to negotiate. In the early 1990s, BP sent a team to investigate a prospective gas field in East Siberia called Kovykta, to which it had first been alerted by local Russian environmental activists.[40] Over the next decade, BP invested heavily in exploring the field. Its vision was that it could build a pipeline passing through Mongolia to supply East Siberian gas to Beijing. But BP was ahead of its time. The Chinese gas sector in those days was an appendage of the oil industry; it was treated as a poor cousin and hardly figured on the Chinese leaders' radar screen. During a decade of fruitless talks, BP lost control of the license to the Kovykta field and exited from the project without having anything to show for it.[41]

Thirty years later, in January 2020, BP's vision was finally realized, with the inauguration of the Power of Siberia pipeline. But by this time the project had completely changed face: Gazprom was now the Russian company in charge, and its sole partner was CNPC, a Chinese state-owned company. The Kovykta field had been paired with a second field in the Russian Far East called Chayanda, and it now formed the anchor of a vast pipeline system that looped 3,000 kilometers across the map in a circle to China's backward northeast province of Heilungchiang. The Power of Siberia pipeline is now the linchpin of Russia's "pivot to the East," a policy promoted by the Russian state with the vigorous support of President Putin, aimed at developing a privileged economic and political relationship with China.[42]

The potential market is huge. China is expected to increase its gas consumption from 300 bcm per year to more than 500 bcm in 2030. But Russian gas to China will face growing competition from every source. Turkmen gas (33 bcm to China in 2018) was joined by gas from Kazakhstan in 2017, and Central Asian volumes are expected to increase steadily. Central Asian gas is being developed through tied credits with Chinese companies, under which the Central Asians receive materials and labor from China, and Russian strategists are pessimistic that Russian gas will be able to compete with it.[43]

From the beginning Gazprom has promoted a second route to China, from West Siberia through the Altai gap between Kazakhstan and Mongolia to the western end of the West-East pipeline. The advantages of the Altai route would be considerable for Gazprom: the distance is shorter than the Power of Siberia, and it draws from the abundant reserves now available to Gazprom since the development of the

Yamal Peninsula. The Chinese have so far resisted the Altai route, even though it has been vigorously endorsed by President Putin. But the Russian side remains hopeful, even reviving BP's old proposal to transit gas through Mongolia, but this time from West Siberia. Gazprom's persistence is understandable, in light of the possible decline of Russian pipeline exports to Europe and the anticipated continued strong growth of Chinese gas demand.[44] But the Russians know from long experience that striking deals with the Chinese takes patience. Elena Burmistrova, the deputy chair of Gazprom and the head of Gazprom Export, says philosophically:

> Because negotiations on the first Power of Siberia project did in fact last around 10–15 years, these talks had many different phases—some more [active], which, it seemed, were bringing us close to an agreement, then we would again go our separate ways for a year or two, so on and so forth. That is, with the Chinese, all plans are made for a century, as you know. But, nevertheless, we continue to discuss the Western route and possible shipments from the Russian Far East.[45]

Chinese coal remains the great competitor to Russian gas in China, as Russian officials are well aware.[46] As it happens, much of the recent Chinese growth in demand for coal has been concentrated in the northern regions of China, through which Russian pipelines must pass. Russian pipeline gas also faces competition from LNG, including its own. As shipments through the Power of Siberia pipeline build up, it will take up to a decade for Russian pipeline gas to proceed down the Chinese coast, displacing coal as it goes. Meanwhile, LNG will be expanding in the opposite direction, from the southeast, where most of the country's gas-fired capacity is located. This will now include Russian LNG; the first Russian LNG exports from Yamal to China began in 2019, with 4 bcm.[47] But Russia has a lot of catching up to do. China imported 54 mt of LNG in 2018; of this total, Australia accounted for 42 percent, followed by Qatar, Malaysia, and Indonesia. Russia did not figure on the list.[48]

Russian strategists speak optimistically of future Russian exports to China of as much as 100 billion cubic meters per year (bcma), but this assumes that all of the projects in Gazprom's wish list will be realized, and indeed that the Power of Siberia pipeline will ramp up quickly to its rated 38 bcma. Considering that it has taken three decades to reach this point, progress to 2050 may well fall short of the Russians' 100 bcma target.[49]

Meanwhile, other future export markets beckon—at least in principle—but Russia is a long way from being able to take advantage of them. India is the prime example. India's coal industry and its coal-fired power fleet are perennial loss-makers, partly as a result of the low quality of Indian coal, and partly because of the poor state of the rail-roads. As a result, even its new coal-fired power plants are idle half the time. India has ambitious plans to adopt LNG, and it is already the world's fourth-largest LNG buyer, but it is short on regasification capacity, and its gas transmission system, at a mere 17,000 kilometers in 2020, is woefully inadequate for distribution throughout the country (although ironically, at present one-half of it is underutilized). Global major players are stepping up to supply LNG to India, notably Shell and ExxonMobil, with recent alliances to promote LNG in transportation.[50] But significantly, Russian LNG has played virtually no role to date, apart from the delivery of occasional cargoes by Gazprom Export. India has held talks with both Novatek and Rosneft about investing in their Arctic projects, but at this writing no firm deals have been concluded, and so far this has not resulted in deliveries to India. This market will doubtless take some time to develop from its present embryonic state.[51]

Embodied Gas as a Growing Export?

Russian energy strategists have been giving growing attention to the potential for exports of "embodied gas"—products that are made with gas and can then be exported, such as fertilizers, some metals, and gas-based chemicals. Historically the petrochemicals industry in Russia was based primarily on oil, not gas, and only recently has the attention of policymakers begun to shift. If there is an overabundance of gas supply capacity, but constraints on gas exports, how much could be realized from exports of embodied gas?[52]

A decade ago the outlook seemed especially bright in the area of polymers, the basic raw materials for plastics and films. The petrochemicals industry planned construction of six major clusters to double the production of polymers by 2030, with the aim of reaching the same per capita domestic consumption as in Europe, as well as boosting exports.[53] However, subsequent developments have forced a scaling down. First was the economic slowdown in Russia, which cut domestic consumption of polymer products. Second was the rapid buildup of polymer production in China, which displaced Russian exports not only to China but also to Europe. Russian polymers were too expensive and too poorly marketed to compete.[54] This has negative implications for several of Russia's petrochemical clusters. The Sayankhimplast cluster, for example, is intended to provide an outlet for liquids produced in association with gas from the Kovykta field, one of the sources for the Power of Siberia pipeline to China; similarly, the Amur gas-processing cluster uses gas liquids from the Chayanda field. Any delays in the development of these clusters not only sets back plans for the regional development of East Siberia and the Far East, but also weakens the economics of the Power of Siberia pipeline.[55]

The biggest single gas chemicals project on the drawing board is a joint venture between Gazprom and a newly formed company called Rusgazdobycha, owned by interests connected to a prominent family of oligarchs. The planned complex, to be located at Ust'-Luga at the western terminus of the Yamal Pipeline near Saint Petersburg, is designed to produce LNG as well as feedstock for polymers. It would be Gazprom's first entry into the LNG business in Europe, but the primary purpose of the project is to extract liquids as feedstock for chemicals.[56]

In addition, Russian companies have ambitious plans for using natural gas to produce methanol and ammonia. These are much more tentative, however, because of the stagnation of the European market for these products, as well as the recent vigorous growth in domestic production of gas-derived chemicals in China, which hitherto has been a major importer.[57]

Russian analysts estimate that if Russia's plans for gas chemicals come to fruition, they could generate as much as $7–8 billion per year of additional export revenue. However, critics complain of a lack of capital and poor coordination, and the outlook is presently uncertain. In sum, Russia cannot expect salvation from gas chemicals.

What other alternatives might be available? Suddenly, all the talk is about hydrogen.

Hydrogen: The New Holy Grail of the Energy Transition

In the last few years, hydrogen swept the world as the great new hope of climate-change policy. Eight countries have adopted national hydrogen strategies, beginning with Japan in 2017 and followed by New Zealand and Australia in 2019. But the excitement over hydrogen has been most pronounced in Europe, where the Netherlands, Denmark, Portugal, Norway, and Germany all adopted national strategies in 2020, culminating with the European Union's strategy the same year. These countries are betting that hydrogen can become the backbone of European decarbonization policy; but as we shall see, these hopes are still a long way from being realized.

Hydrogen can be produced in several ways, which are referred to by various colors.[58] The most important ones are "blue," "turquoise," and "green." Blue hydrogen is made using steam (so-called steam reforming) to produce hydrogen from gas or coal; this process also yields CO_2 as a by-product, which is then permanently stored underground. Turquoise hydrogen is made by splitting natural gas at high temperatures into hydrogen and solid carbon (a process called pyrolysis). Green hydrogen is generated by electrolysis from renewable energy. All three methods hold the promise of being carbon-neutral, but the disadvantage of the blue process is that there are only limited opportunities for underground storage, and such storage is not guaranteed to be permanent. Therefore, the turquoise and green methods have attracted the most recent attention. Turquoise hydrogen, because it is made directly from natural gas, would provide a carbon-neutral means of exploiting the world's vast reserves of natural gas, given that the solid carbon it produces can be put to other, nonpolluting uses. Green hydrogen would bypass natural gas altogether, using the world's increasingly abundant supplies of renewable energy. Not surprisingly, gas producers favor turquoise hydrogen because they could continue producing gas. Environmentalists oppose it for the same reason; they favor green hydrogen as the most promising decarbonization technology of the future. Yet there is a strong case for using blue and turquoise hydrogen as transition fuels until green hydrogen can be deployed on a large scale. This is the compromise approach embodied in the EU strategy, and in the German and Dutch programs.

Hydrogen has many advantages as a fuel. It is the most abundant element in the universe, and on Earth it is not confined to limited regions or deposits. Although it is not itself a primary energy source, once created it burns cleanly, and when generated by electrolysis from renewable electricity, it causes no pollution. But hydrogen is not without its faults. It is less energy-dense than natural gas, and therefore must be compressed or liquefied, at a cost in energy. It is highly explosive, which makes it tricky to store and to transport. It tends to leak, because of the small size of the molecule. It is inefficient to ship by train or truck, because of its low density. It is also corrosive; in particular, it degrades (decarbonizes) steel, which makes it impossible to ship pure hydrogen by today's pipelines and distribution systems.

But the most fundamental problem with hydrogen is that, at present, and probably for some years to come, turquoise and green hydrogen have yet to be proven economical, and will not be available for many years at the scale required. In addition, hydrogen suffers from a chicken-and-egg problem: it is currently too costly to tempt potential customers to use it on the necessary scale, yet large-scale demand is essential if costs are to be brought down. To achieve this, large-scale infrastructure investments will be needed to connect producers to users, yet there is as yet no answer as to who will foot the bill.

At first sight this is a puzzle. Why have politicians and policymakers, especially in Europe, raced ahead of the technology and the economics? The answer is that hydrogen would be the ideal answer to a number of problems currently confronting energy-transition policy. Climate-change policy in Europe is currently largely focused on power; but achieving zero-carbon targets will require solutions in manufacturing, transportation, construction, and heating. That is where hydrogen comes in. If the many problems above can be solved, hydrogen would become the key to decarbonization.

Hydrogen in Russia: Talk and Talk?

Hydrogen has suddenly burst on the policy scene in Russia as well. Just since 2019 Moscow has witnessed an explosion of conferences, road maps, and dozens of media articles devoted to hydrogen.[59] "But the surprising thing," comments Tat'iana Mitrova, the head of the Skolkovo Energy Center, "is that even so no pilot projects have appeared, only tons of paper. . . . We endlessly talk and talk."[60] Will Russian hydrogen become real?

Not surprisingly, much of Russia's newfound interest in hydrogen comes from Gazprom.[61] Hydrogen is both a threat and a possible opportunity for Russia's gas business in Europe. A threat because if Europe succeeds in developing green hydrogen, produced with re-newables, as an alternative to gas, that could close much of the Euro-pean gas market to imported gas. But hydrogen could also be an op-portunity to preserve that market, if it can be mixed with gas in Europe's gas infrastructure. In particular, Gazprom hopes that its vast export pipeline system can be used—if Russia can produce the nec-essary volume of hydrogen.[62] Blending limits, though, as noted, could impose an upper bound on how much hydrogen could be exported together with gas. But it is still uncertain what proportions of hydrogen to gas are safe or economical.[63]

Germany, in turn, has taken note of the growing Russian interest in hydrogen. In 2019 Gazprom and the German utility Uniper began discussing possible collaboration on hydrogen.[64] In 2020 Gazprom's subsidiary Energieholding was in talks with Siemens to jointly develop a gas turbine adapted for hydrogen.[65] It is possible that hydrogen could turn into a new basis for energy collaboration between Germany and Russia.

In parallel, Gazprom has also gone on the offensive to defend its existing business. At the European Gas Conference in Paris in November 2019 (one of the last conferences the delegates were able to attend in person before the onset of COVID-19), the head of Gazprom Export, Elena Burmistrova, warned bluntly that the "demonization" of gas by climate-change radicals and an overhasty commitment to renewables and hydrogen would destabilize the European economy and "destroy all its advantages." The oil and gas companies are already spending billions of dollars on renewables and hydrogen, she said, but dependence on fossil fuels will not vanish overnight. She did not reject the EU's commitment to net-zero carbon by 2050; on the contrary, she said that "by that time we will have adapted our business to the new realities."[66] But in the meantime, Gazprom means to continue selling gas.

One could choose to be cynical about the Russians' recent response to hydrogen. It fits squarely into the framework described in Chapter 1, as corporate players respond to threats arising from outside Russia with green advertisements and initiatives. The alarm in Russia is real. The question is how imminent the threat is.

There are a number of possible outcomes. Green hydrogen is unlikely to be a major factor any time soon because it will be more cost-effective to put power from renewables straight into the power grid, leaving too little renewable power to make green hydrogen on any significant scale. Consequently, if hydrogen is to make inroads over the next decade or two, it will need to be blue hydrogen. One avenue might be to generate hydrogen in Russia, storing the resulting CO_2 in old oilfields, as Norway is doing, and to ship the hydrogen to Europe.

Another might be to continue shipping gas and to turn it into hydrogen at the destination. All of these solutions will require experimentation, investment, and time. As a result, gas demand could hold up in Europe longer than currently expected.

Conclusions: Will Russian Gas Replace Oil?

The chief effect of climate change on Russian gas will be to constrain Russia's gas exports and the profits from them, despite the anticipated worldwide boom in gas demand.

There will be little direct impact of climate change on Russian gas inside Russia itself, and there will be little domestic backlash against it. In this respect Russia is a different world from the United States. Even though both have a virtually unlimited abundance of cheap gas, Russian gas is mostly conventional and is located in remote areas, whereas much of US gas is unconventional shale gas, extracted by fracking, and a significant part of it comes from heavily populated areas. As a result, US gas production is vulnerable to popular opposition to fracking, whereas Russian production is not—at least not to opposition at home. Even the oil industry's flaring of associated gas is not a public issue because the policy of limiting gas flaring has come from the government, not from the population. In sum, climate change per se will have little direct impact inside Russia on gas production.

The carbon footprint of gas will likewise not affect Russia's domestic gas demand. Gas is still viewed in Russia as a virtuous fuel, not least by Putin himself, but public opinion does not disagree. Russia is already largely a gas-fired economy. Gazprom's program to extend gas use in villages, called *gazifikatsiia,* is popular, as is its program to displace coal in the Far East. There will be little public reaction to Russia's new petrochemical clusters, which are located in remote areas with small populations. In the eyes of the population, gas remains largely the invisible benefactor.

It is in Russia's export markets that the impact of climate change will be felt. The European gas market is headed for decline, although whether rapidly or slowly will depend on the rate of growth of renewables, the availability of alternatives such as green hydrogen, and possible political developments such as the adoption of a carbon border tax in Europe (discussed in Chapter 1). The other big question ahead is whether Russian gas, chiefly in the form of LNG, will be able to compete profitably in the Chinese market. The competition from Chinese coal will be intense for the remainder of the 2020s. Thereafter, Russian LNG will face fierce competition in Asia from numerous ambitious suppliers, such as Australia and Qatar and possibly Mozambique.[67] Russia will be fortunate to reach its planned 20 percent share of the world LNG market. In addition, LNG buyers, especially China, will enjoy strong bargaining leverage, which will keep prices and profits down.

All of these factors add up to a pessimistic long-term outlook for Russian gas. The combination of lower export volumes by mid-century, together with lower global gas prices (in today's dollars) means that Russia's future gas revenues will not come close to replacing those from oil. In sum, as oil revenues decline, gas may help, but it cannot come to the rescue.

4

RUSSIA'S COAL DILEMMA

In the winter of 2018, local residents of Kemerovo, the capital city of the Kuzbass coal region, rejoiced as fresh snow blanketed the soot and ash that normally cover it. Or so it seemed. But when they went out to make snowballs, they discovered that the "snow" was actually white paint that had been sprayed over the black sludge. The mayor of the city denied that he had ordered the whitewash, and some lower-ranking officials were reprimanded. The next snowfall was black, as usual.[1]

The white paint episode captures something of the atmosphere of Kemerovo. The town is entirely devoted to coal, and its inhabitants depend wholly on the coal mines. The province's governor, Sergei Tsivilev, is a businessman-turned-politician who has a gift for the popular touch, such as when he knelt in front a crowd of protesters after a fire in a mall had killed sixty-four people, to beg their forgiveness.[2] He is a vigorous promoter of coal. He maintains a daily video blog, on which he talks up the prospects for the coal sector and lambastes opponents of the coal industry. One of his favorite targets is the railroads, which he regards as insufficiently committed to coal. Recently he complained that 23 million tons (mt) of Kuzbass coal, destined for export to the east, were sitting on site in the Kuzbass, for lack of railroad capacity.[3]

Like other Russian industry leaders we have met in this book, Tsivilev arrived at his present post by a winding path. He started out as a businessman / investor in Saint Petersburg, with ties to well-connected partners such as Gennady Timchenko, an intimate of Putin's circle. Through these connections, Tsivilev made his first entry into the coal business in 2010, as co-founder, along with Timchenko, of the Kolmar coal company, which produces coal in Sakha; in 2014 he became CEO. The key event in Tsivilev's career was a meeting with Vladimir Putin in February 2018; shortly afterward Putin named him interim governor of Kemerovo, replacing the longtime governor Aman Tuleev. He was confirmed in the post when Putin visited Kemerovo in 2018.

But though a recent recruit to the coal business, Tsivilev has become one of the most important defenders of the sector. For Tsivilev, climate change hardly exists and coal remains the future. Putin is one of his biggest supporters. As we shall see in this chapter, the coal industry and politicians in coal-mining regions have been the most conservative force in Russia's climate politics, opposing any initiative that might impose costs on the industry or interfere with its prospects. Tsivilev can relax: the Russian coal industry is thriving. Yet for how long?

Coal—the Zombie Fossil Fuel

Coal is the leading source of greenhouse gas emissions in the world, and emissions from coal are still growing. In 2019, out of a total of 34.17 gigatons of CO_2 emissions from energy consumption,[4] coal produced about 45 percent, up from about 40.5 percent in 1990. Because this share is measured from a far smaller base, the absolute increase in annual coal-based emissions has been a massive 12.7 gigatons, or a 62 percent rise in thirty years.[5]

Coal is still the second-leading source of primary energy consumption in the world, with a 27 percent share in 2019, just behind oil. Decade after decade, it continues to fuel nearly 40 percent of the world's generation of electricity.[6] For Russia, it is a rising source of export revenues, thanks to a vigorous reorientation of its coal industry toward exports. The coal sector, more than any other, underscores the contradictions in Russia's energy signature. The coal industry and politicians from coal-producing regions resist the climate-change story, oppose measures to deal with it, and remain focused on increasing coal production, primarily for export. President Putin, despite his claim that Russia has the greenest energy exports in the world, has been wholly supportive of coal's continued expansion.

Russian coal, indeed, has been one of the leading success stories of the Putin era. The coal industry and the government have spent massive amounts over the last thirty years to revive and modernize the coal sector. The key to its success has been a reorientation from domestic consumption to exports. By comparison with domestic consumption, which has been stagnant, Russian coal exports have been booming.[7]

To put this story in perspective, however, coal occupies only a distant fifth place as a source of Russian export revenues. In 2019 (the last normal year before COVID-19 struck), coal exports made up only 4 percent of total export income, compared to 44 percent for oil and oil products and nearly 12 percent for natural gas and LNG. Exports of metals accounted for 9 percent, more than twice as much as coal (see Chapter 5).[8] Coal exports have a long way to go before they can rival hydrocarbons.

Moreover, coal faces a ticking clock. In coming decades, as climate change accelerates, coal will gradually be reduced to a minor place in the world's fuel balance, or driven out altogether, because of mounting global opposition to pollution and emissions from coal, together with the rapidly improving economics of renewables.[9] Indeed, coal demand may already have peaked. Coal consumption worldwide has declined in four of the last six years. In the member countries of the Organisation for Economic Co-operation and Development (OECD), the share of coal in global primary energy consumption fell to 27 percent in 2019, its lowest level in sixteen years.[10]

Thus, in coming decades Russia will face a shrinking market. The price of coal is likely to decline as well. The long-term outlook for Russian revenues from coal exports is bleak.

There will be little offsetting help from domestic demand. The main sources of Russian coal are located far from most coal-fired power plants and metallurgical smelters, and thus face high transportation

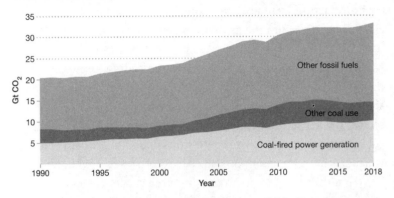

Global energy-related carbon dioxide emissions by source

Source: International Energy Agency, *Global Energy and CO$_2$ Status Report 2019,* https://www.iea.org/reports/global-energy-co2-status-report-2019.

costs. Coal's main competitor in Russia is natural gas, which is cheaper[11] and more conveniently available (see Chapter 3). In coming years the domestic market for coal will shrink to the regions closest to centers of production, mainly southern Siberia and the Far East. The domestic future of Russian coal is to be at best a regional fuel, with a declining share in the overall fuel balance.[12]

However, the global numbers conceal two completely opposite trends: In Europe and the United States, coal is rapidly being squeezed out of the energy mix. But in Asia, economic growth is still pushing coal consumption sharply upward.[13] The main driver is demand for electric power. Supportive state policies and powerful interest groups, together with the "electrification" of overall energy demand, are catalyzing construction of a new generation of coal-fired power plants,[14] which in turn will create pressure to sustain coal demand in Asia tomorrow.

Thus, Russian coal strategy faces two opposing prospects over the next decade: on the one hand, the inexorable shrinkage of the European market, historically Russia's main export destination, combined with stagnation at home; and on the other, the rapid but temporary growth of the Asian market, followed by a slow but terminal decline.

Coal consumption by region

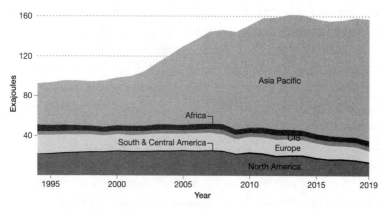

Source: BP, *Statistical Review of World Energy 2020,* https://www.bp.com/en/global/corporate/energy-economics/statistical-review-of-world-energy/coal.html.

As a result, Russia faces a fundamental dilemma—how much to prioritize expanding the coal sector today, if its fate, beginning in another decade, is to fade away?

The Vanishing European Market

More than half of Russia's coal exports currently go to Europe. These consist largely of high-quality steam coal from the Kuznetsk basin (the Kuzbass), most of which is used to generate power. Germany, in particular, still generates more than one-third of its electricity with coal. It imports about 18 mt per year from Russia.[15] In recent years the share of Russian coal in German imports has grown steadily, at the expense of competitors such as Colombia. But Germany is not alone; Russian coal exports to all of Europe have increased steadily over the past decade, particularly to Italy and Spain. Russia enjoys one key advantage: proximity. Over the last decade, the Russian coal companies have reconfigured their operations around exports to Europe, notably by expanding their port capacity on the Baltic coast, and they now dominate the European market.

Yet the European market for Russian coal will not last much beyond another decade because the share of renewables has been steadily increasing and political opposition to coal has been mounting. In Germany, according to the government's recently adopted plan for exit from coal, the last coal-fired power plant will be phased out in 2038, but before that, the capacity of coal-fired plants will be cut from 42.6 gigawatts (GW) in 2017 to 17 GW in 2030, squeezing demand for Russian coal.[16] In parallel, Russian coal has become increasingly unpopular in German public opinion. The Russian coal exporters have come under increasing scrutiny from the German media, both on environmental grounds and on charges of money laundering. German utilities are feeling growing public pressure to cut back on Russian coal.[17] For all of these reasons, Russia's coal exports to Germany appear fated to disappear over the next few years. The same is happening among Russia's other European buyers, chiefly Italy and Spain, although for the time being coal demand remains strong in southeastern Europe, particularly in Turkey. Overall, European demand for coal is expected to decline by 5 percent per year to the mid-2020s,[18] and even faster thereafter.

Russian coal strategists, both in the companies and in the Russian government, are well aware of these trends.[19] As a result, their sights have shifted to the Asian market.

The Booming Asian Coal Market— and the Chinese Wild Card

For the past thirty years, the main cause of the global surge in coal consumption has been China. Largely as a result of the growth in Chinese demand, the Asia-Pacific region, which as recently as 1994 accounted for only about one-quarter of all coal consumption worldwide, now takes up three-quarters. In coal-fired power generation alone, Asia's share soared from just over 20 percent in 1990 to almost 80 percent in 2019, much of that from China.[20]

But when it comes to coal consumption, China is a wild card. In 2016 China adopted new policies aimed at curbing pollution from coal. It restricted coal mining, the use of coal for home heating, and the construction of new coal-fired power plants. It seemed as though the phenomenal boom in Chinese coal demand, which had powered the world coal market for thirty years, was finally at an end. That would have been a game changer for the entire international coal trade.

But as Chinese economic growth slowed, the central government abruptly changed course and issued permits for a whole new generation of coal-fired plants. As of June 2020, nearly 250 GW of coal-fired capacity were under development, of which 98 were under construction and 152 in various stages of planning. The total under development, if realized, would be larger than the entire coal fleets of the United States (246 GW) or India (230 GW).[21] This would put Chinese coal demand on a whole new growth track, with major opportunities for Russian coal.

In 2020, though, China's climate and energy goals shifted again, as Chinese leader Xi Jinping pledged to achieve net zero carbon emissions by 2060. But at this writing, that goal has yet to be translated into lower targets for coal production or power plant construction.[22]

The entire Chinese political system is biased in favor of coal-fired power: state-owned banks lend enthusiastically to state-owned utilities without proper diligence; province governments cheer them on by granting generous lending quotas and calling for more spending on new power plants; and coal-producing provinces lobby for more coal consumption to protect their own production or to avoid importing coal or power from other provinces. In some respects the COVID-19 pandemic has made matters worse, by encouraging the provinces' efforts to offset the pandemic's economic impact.[23]

But as economic growth in China has slowed, the growth rate of Chinese coal demand has slowed as well. This implies that the main driver in coming decades will be Southeast Asia and India, and most analyses (including those of the Russians) are based on that assumption. But even in China, growth has not ceased and may accelerate again in the wake of the COVID-19 pandemic. In sum, coal-fired power generation will continue to rise in the region, driven by economic growth, increasing population, and an expanding middle class using electrical appliances, and backed by entrenched coal interests and supportive government policies.

In short, if there is an expanding future market for Russia coal, it will be in Asia. However, the Asia-Pacific region covers most of its coal consumption with its own production and with vigorous exports from Indonesia and Australia. These are much closer to market than Russian coal, which faces an uphill battle against these local sources, owing primarily (as we shall see in more detail below) to its enormous transportation distances and limited rail capacity. Therefore, the Asia-Pacific market is not necessarily the future bonanza it might seem at first glance.

Moreover, long-term economic trends work against coal in several important Asian markets, notably India, where the largest sources of coal production are in the northeast, while consumption is mainly in the southwest. Indian coal, which consists mainly of low-value brown coal, faces long transportation distances over an unreliable rail network. By the time it reaches its destination, it is too costly to be profitable; as a result, half of India's coal-fired power plants are idle.[24] This could work to the advantage of coal imports. Yet Indian coal continues to enjoy strong domestic political support, and one of the side effects of the COVID-19 pandemic may be to make energy-importing countries more reluctant to depend on outside sources.

How long will the golden age of coal last in Asia? Coal could enjoy a new life ahead, because of a rapid buildup of new coal-fired power capacity in the developing world that preceded the pandemic, driven by abundant domestic and foreign subsidies. In 2019, one study found that over $64 billion per year was being directed to the coal sector worldwide by G20 governments, with more than three-quarters going to support coal-fired power generation. A large fraction of this support consisted of finance for the construction of new coal-fired power plants overseas, with China in particular spending nearly $10 billion per year for coal-burning power plants outside China, over and above its own domestic program.[25]

Yet within the power sector, the rapidly changing economics of renewables and storage are shifting the cost equation away from coal, even in Asia. Until recently the total cost (i.e., levelized cost) of a coal-fired power plant was lower than that of a solar or wind plant. But the costs of renewables and storage are dropping so fast that coal is increasingly unable to compete. According to a recent study, "Solar PV and onshore wind are now the cheapest sources of new-build generation for at least two-thirds of the global population." In many places the total costs of a new renewable power plant are lower than the current operating costs of an existing coal-fired one. This trend will strengthen as the decades go by, resulting in abandoned coal-fired capacity and a steady decline in electricity output from coal, in Asia as elsewhere.[26] The golden age of coal in Asia might last no more than another decade.

In sum, the international coal market is likely to evolve in two phases. Until about 2030, global coal demand will continue to grow, chiefly driven by coal-fired generation in Southeast Asia, while European demand gradually fades. Beyond 2030, however, the inexorable growth of renewables, combined with growing opposition to pollution from coal, will begin to drive Asian coal demand down as well.

This implies that there must also be two phases to Russian coal strategy: during the first phase, Russia must adapt its production and transportation systems to the shift from Europe to Asia. During the second, Russia must reduce its costs to enable it to compete with its rivals in Asia, chiefly Australia, and to maintain market share in a shrinking market. But the greatest challenge of all will come at midcentury and after, when, owing to climate change, Russia's coal export market will largely vanish altogether.

Not surprisingly, that is a narrative the Russian coal industry does not accept. The one it prefers to focus on instead—Governor Tsivilev's favorite narrative—is its own remarkable transformation in the last three decades, the story of its revival from the ruins of the post-Soviet collapse and its emergence as a major force in the global market.

The Renaissance of Russian Coal

Until the 1960s the Soviet Union was a coal-fired economy. Then the discovery of giant oilfields, first in the Volga-Urals basin and then in West Siberia, pushed coal into the background in favor of oil and then gas. By 1990, at the end of the Soviet period, coal accounted for only 20 percent of total primary energy production.[27] Moreover, it was a chronically loss-making sector, as it had been Soviet policy to underprice coal to subsidize home heating, power, and industry, and the coal sector was virtually left to fend for itself. As a result, the sector was undercapitalized, obsolete, and woefully inefficient.[28] It was increasingly based on poor-quality lignites—soft brown coals with low heat content—and it faced extraordinary transportation distances and an overburdened railroad structure. Most of the country's coal-fired fleet was more than a generation old. The 1990s only made matters worse.

The renaissance of Russia's coal sector after 2000 is all the more remarkable. It occurred due to a conjunction of several things: a radical restructuring of the industry (which included over $500 million in loans from the World Bank), privatization, investment in new technologies, and above all, as noted, a massive reorientation from the domestic market to exports.

As a result, today's coal industry is unrecognizable compared to that of twenty years ago.[29] Older mines were closed; marginal regions were taken out of production; half of the previous workforce was shed; and direct subsidies were largely eliminated (although, as we shall see, indirect subsidies remain important). The previous state monopoly that controlled the industry was abolished. The locus of the industry has moved east to Siberia and the Russian Far East. The industry has been extensively modernized and today relies largely on open-pit mines, which produce three-quarters of all Russian coal. The strategic center of the industry today is the Kuzbass, which is located about equally far from Europe and the Far East, making it a key "swing producer" that can export both east and west depending on market conditions. Its main products today are high-quality steam coal for power generation and coking coal used in metallurgy, both of which are prized in foreign markets.

The renaissance of the coal industry continued strongly throughout the decade of the 2010s. Between 2010 and 2019, coal production grew by 30 percent, setting a record on the eve of the COVID-19 pandemic at 441 mt.[30] This was propelled by vigorous investment, which increased by 250 percent from 2008 to 2018.[31]

The coal industry has been reorganized into giant vertically integrated complexes, which control power plants, steel plants, railroads, and port facilities. Thus, the entire coal sector today is highly integrated—with the key exception of the rail system itself, which remains controlled by the state-owned railroad monopoly RZhD (Rossiiskie Zheleznye Dorogi). It has been massively reoriented toward exports. In 2008 only one-third of Russian coal was exported,[32] but by 2018 the share had grown to over one-half, and it was still growing.[33] One of the coal companies' main investment priorities has been the construction of new coal-export terminals, which can now handle three times as much traffic as a decade ago.[34] But rail capacity, to connect the mines to the new terminals, remains a severe bottleneck.

Russian Coal Policy

During the turbulent 1990s and early 2000s, coal was treated primarily as a regional fuel. Until recently it was only a minor earner of export revenue, and unlike the oil and gas industries, it was not central to the state's revenues. Coal is a high-cost resource, especially once transportation is reckoned in, and under the current pricing system it cannot compete with gas in most of the country. Perhaps for these reasons, coal gets less of the government's attention, and less of its money in subsidies, than other parts of the energy sector, as we will see below.

Coal, as noted earlier, has little future inside Russia's domestic market. The share of coal-fired generation in Russia peaked in 2000 at 20 percent, and it has been declining steadily ever since.[35] The main reason is competition with natural gas, which now fuels more than half of all power generation. In the western two-thirds of the country, the low price of gas, combined with coal's high transportation costs, excludes coal from the power market. Coal's last remaining redoubts are in Siberia, where the share of coal is still almost 50 percent, and the Russian Far East, where it is 42 percent. Yet even in those areas coal is under challenge, as Gazprom is invading coal territory with a vigorous program of pipeline construction. As Gazprom's pipeline network expands, coal retreats further.

Unlike the coal industry itself, which has undergone considerable modernization, the technological level of Russia's coal-fired power plants is still mostly that of the 1950s. There has been almost no construction of new coal-fired plants since the end of the Soviet Union, and one-third of Russia's coal-fired plants are more than fifty years old. (Moscow's last coal-fired plant was built in 1960.) Unlike the Japanese and the Chinese, the Russians have made no progress in "clean coal" and make virtually no use of supercritical technology.[36]

State spending for coal is clearly dwarfed by the immense amount of state money made available to the oil and gas industries. In 2015 a review of energy subsidies worldwide by the G20 countries concluded that Russian oil and gas companies and projects received an annual average of $22.8 billion in national subsidies, while the coal industry received only $43 million. In reality, the coal industry gets more than that, but the bulk of the state's funding consists of assistance to cover the closing of mines, the support of abandoned mining towns, and the relocation of miners.[37] Much of the state's support for coal has taken the form of loans through the state-owned banks VEB.RF, VTB, and Gazprombank, mainly for the purpose of promoting steel. But these have not played a major role in the revival of coal industry, which has been financed largely by the private coal-and-steel conglomerates themselves.[38]

Where state funding is crucial, however, is in rail transportation, because of its impact on exports. The Russian railroads and the coal industry are acutely dependent on one another. The coal industry accounts for nearly half of all freight traffic on the rail system, and in some places, like the Kuzbass, the figure is closer to 85 percent. There is ample rail capacity on the westward routes to Europe, but on the eastward routes to Asia there are serious bottlenecks that are holding back the growth of Russian coal export capacity. The government has launched an ambitious program, called Eastern Polygon, to expand the capacity of the two main rail arteries, the Trans-Siberian Railway (Transsib) and the Baikal–Amur Mainline (BAM).[39] In his "May Ukaz" of May 2018, Putin laid down ambitious targets for the expansion of the coal-carrying capacity of the Transsib and the BAM by laying 1,310 kilometers of new track, at a cost of 490 billion rubles (then $7.7 billion), to raise the system's capacity to 180 mt per year of coal by 2024.[40] President Putin has made this one of his national projects, and he has spoken out repeatedly to urge rapid progress.

However, the coal companies, together with their regional supporters such as Governor Tsivilev, complain that the program is running behind plan, and that the president's personal interventions have had little effect. A glance at the webpage of RZhD does raise questions. On the one hand, the expansion of the main eastward arteries—the BAM and the Transsib—is listed as the company's top priority, at the head of the "president's list" of investment projects. Yet while RZhD's overall investments increased strongly between 2015 and 2019, from 365 billion rubles (then about $6 billion) to 674 billion rubles (then about $10.7 billion), the amount actually spent on the BAM-Transsib expansion has declined sharply from a peak of 84 billion rubles (then about $1.3 billion) in 2016 to only 35 billion rubles (then about $0.5 billion) in 2019.[41] However, the complaints of the coal industry did have an effect: in mid-September 2020 the Russian media reported that RZhD had sharply raised its spending plans for the Eastern

Polygon.[42] But it remains to be seen whether this trend will continue in the face of the COVID-19 pandemic.

Climate Change and Russia's Coal Strategy

How does the Russian coal industry see its own future, in the face of the growing challenge of climate change? In public, it mostly denies that a problem exists. As we saw in Chapter 1, the coal industry joined together with other branches of heavy industry to fight Russia's ratification of the Paris Agreement. The debates on the floor of the Duma were fierce. Several deputies, including those from coal-mining regions, argued that climate change was fake and that the Paris Agreement was a plot directed against developing countries. Similarly, in industry groups such as the Russian Council of Industrialists and Producers, representatives of the coal industry have taken strong public positions against any measures to monitor and control CO_2 emissions.[43]

But internally the leading coal companies increasingly accept that climate change is real, and they are mapping their outlook accordingly. A good example is SUEK, Russia's leading coal company. Its chief strategist, Vladimir Tuzov, comes to his job with extensive experience in the West.[44] For Tuzov and the younger generation of strategists like him,[45] the debate over climate change is over; the question is how to find opportunity in it. And the chief opportunity, as we have seen, is in Asia.[46]

Tuzov's reasoning goes like this: The international coal trade accounts for about one-quarter of global coal production. That share is not expected to grow much over the next decade, perhaps 10 percent by 2030. But within that stable envelope there will be major demand growth from Southeast Asia, chiefly from Vietnam, Bangladesh, Malaysia, and the Philippines. Demand from India will grow strongly as well, despite the Indian government's attempts to rein it in. In these areas the chief competition will be from Australia. In contrast, Indonesian exports, which are formidable today, will soon fade, because growing internal demand will absorb most of its production.

But for Tuzov and SUEK, China is key. First, even though its economic growth is slowing down, China will continue to import seaborne coal, because its major consumption is located close to the coast. Second, its ambitious program of financing and building coal-fired power plants throughout the developing world will provide growing demand. Lastly, the accelerating boom in electric vehicles in China will require large amounts of additional electricity. Over the next decade this will be supplied mainly by coal-fired plants that are currently operating at only half capacity. In short, Tuzov concludes with a smile, the rise of green transportation in Asia promises a bright future for the Russian coal industry.

But the optimism of the coal industry is not shared by all the Russian players, and in particular by the Energy Ministry. In its 2019 *Energy Strategy to 2035* (discussed in Chapter 2), the Energy Ministry warned of the many obstacles facing the coal industry. Its analysis essentially parallels that of this chapter. It warns of declining domestic demand, and it is also keenly aware of the potential impact of Chinese restrictions on the mining and use of coal. Even Anatoly Yanovsky, the long-serving deputy minister of energy who is responsible for monitoring coal exports and might have been expected to talk up the industry's prospects, warns instead that Russian coal exports to China could be limited.[47]

The coal industry, at least in public, dismisses these fears. SUEK's Tuzov projects that Russian coal exports have the potential to reach 296 mt in 2030 compared to 236 mt in 2018, a 25 percent increase.[48] During that same period Russia's competitors—chiefly Australia, Colombia, and South Africa—will increase their export potential by only about 25 mt per year. In the Asia-Pacific market, the gap between supply and demand could widen to 220 mt per year. For Tuzov, the prize is Russia's for the taking. By 2030, he argues, Russia could hold as much as a quarter of the world's coal market.

This optimistic projection assumes that the Russian coal companies will be able to increase production sufficiently to generate such a large export surplus. This will require them to continue investing heavily while keeping their costs under control. Only about one-quarter of the expected increase will come from the Kuzbass and West Siberia, where production will soon reach its limit; the remainder will be from East Siberia and the Far East. This is good news because it will lessen the need for new west-to-east rail capacity, but expanding production in the east will require considerable modernization and expansion of coal production in places like Sakha and Amur, where the industry is still only weakly developed.

But this scenario only extends to 2030. Will Russia after 2030 find itself in the position of having the world's largest export capacity but only a declining market for it?

At present there is little public recognition of this dilemma among Russian analysts and policymakers. Putin does not address it, apart from ritual references to "environmental quality." In the government's official *Energy Strategy to 2035*, at the end of a long list of risks facing the coal industry, there is only a brief reference to the "international campaign against coal under the heading of realization of the ecological agenda."[49] One of the few personalities to address the issue head-on is Tat'iana Lan'shina, a member of the Russian delegation to the 2019 COP25 conference in Madrid, who declares flatly, "There won't be anyone to sell it to. Coal is economically dead. Everyone understands that."[50] But that is clearly not yet the understanding in Russia as a whole, where both policymakers and business leaders project continued growth ahead.

Yet if the "death of coal" narrative comes true by mid-century, what would be the implications for Russia's export revenues? Export prices for Russian coal have been highly unpredictable, ranging from $31.3 per ton in 2000 to a high of $94 in 2017. During that period export revenues grew steadily, from $9 billion in 2010 to $17 billion in 2018.[51] But Russia's future revenues from coal exports will depend on the balance between the coming decline in Europe and the expected continued growth in Asia. To take the key case of China, Chinese imports of Russian coal would have to double from their present level to make up for the disappearance of European imports.[52] To generate net growth, countries like India and Taiwan would have to increase their imports of Russian coal substantially above their present levels of 8 mt apiece. This is not impossible, but it depends on uncertainties such as the unpredictability of Chinese coal policy, the penetration of renewables, and the opposition to pollution from coal against the political resistance of established coal consumers in Asia.

Taking these factors into account, the Russian coal industry's projections look optimistic. It is more likely that Russia's revenues from coal, after a brief boom in the 2020s, will at best end up at about their present level, and will more likely decline. The snow in Kemerovo may be whiter as a result, but that will not be good news for the Kuzbass.

In this chapter and the last two we have analyzed the prospects for Russia's three leading fossil fuel exports—oil, gas, and coal—which together account today for 60 percent of Russia's export revenues. By mid-century, all three are likely to be declining. What alternatives does Russia have? In Chapter 5 we turn to the one energy source that is sweeping markets all over the world and ultimately threatens the final demise of coal—renewables.

5

RENEWABLES

A Slow Start

In a packed lecture room in Moscow's elite engineering school, the renowned Bauman Moscow State Technical University, several hundred students are listening attentively to the middle-aged man at the podium. A little stouter now than he once was, he has lost nothing of his trademark rust-red hair and his self-assurance. He was famous before they were born, when he was known as the father of Russian privatization.[1] Today he is the most eloquent voice in Russia for the future of renewables. His name is Anatoly Chubais.

Over the past forty years, Chubais has been the most significant figure in the history of Russia's post-Soviet transformation into a market economy. Many readers will find that a controversial statement. In the 1990s, while still in his thirties, Chubais led the mass privatization of Russian industry and commerce by distributing free shares to the population—an accomplishment that was overshadowed by the infamous "shares for loans" scandal that led to a giveaway of some of the most valuable assets of the country to the future oligarchs, permanently tarnishing his reputation with the Russian public.[2]

But in the 2000s a more cautious Chubais reinvented himself as the head of the Russian electricity monopoly, United Power Systems (UES), where he fought—and largely won—a bitter decade-long battle for the radical restructuring of the power sector.[3] Then, in his mid-fifties, he founded Rusnano, a visionary new state-sponsored start-up charged with developing next-generation nanotechnologies such as graphene, a revolutionary new form of carbon with many potential applications.[4] Now in his mid-sixties, his innovative verve is undiminished, and he has lately reinvented himself yet again, as the would-be father of Russian sun and wind.[5] In December 2020 he left Rusnano and was named Putin's special representative to international organizations, a role that may keep him at the center of Russia's climate negotiations.[6] At every point, for better and for worse—admire him or dislike him—Chubais has led the way for his view of Russia's future.

Armed with PowerPoint slides that are reminiscent of Al Gore's Nobel Prize–winning film *An Inconvenient Truth*, Chubais travels the country, preaching the virtues of renewables and their potential for Russia. It's not that Russians don't believe in renewables, he says; rather, the problem is that they are uninformed, and there are many myths. One of them is that Russia is too dark and too cold for solar, and that in much of the country the population is too sparse. But the city of Cheliabinsk in the Urals, Chubais shows his audiences, gets the same amount of sunshine as Berlin, and solar panels work better the colder it gets. As for wind power, he points out, Russia with its 17 million square kilometers of land mass, is potentially the world's leader, especially in the Far North, where wind power is especially well suited to Siberia's isolated communities[7] (some 60 percent of Russia's territory has no access to power from a grid).[8] Chubais does more than talk: during his twelve years at the head of Rusnano he invested seed money in Russia's first plants to produce solar cells and blades for windmills (see more on this below). During that time, he was Russia's leading renewables entrepreneur.

As with nearly everything in Russia, the state is never far away. There is a state plan for the development of renewables.[9] There is state financing, if on a limited scale; Rusnano itself was created with seed money from the state, and other state-owned companies like Rosatom, the Russian nuclear giant, are now getting involved. Chubais tells anyone who will listen that no one in the world can develop renewables without state support—and Chubais has perhaps more expertise in dealing with state bureaucracy than anyone in Russia. His audience even includes Vladimir Putin. In March 2019 Chubais briefed Putin on renewables at the Kremlin, reporting on his progress and informing the president that he had the first export contract for a Russian-made solar panel, to Kazakhstan. Putin nodded approvingly, although nine months later he dismissed Chubais and folded Rusnano into a new state-owned conglomerate under the control of VEB bank.[10]

Yet despite the efforts of Chubais and a handful of like-minded visionaries and entrepreneurs such as Victor Veksel'berg's Renova, renewables play only a very limited role in Russia's landscape, and that is unlikely to change. Russian renewables face fundamental obstacles, stemming chiefly from the country's unique endowment in hydrocarbons, especially gas, and the nearly impregnable position of gas in the country's political and regulatory system, the result of the domestic market for gas analyzed in Chapter 4. The potential for exports of renewable technology seems slim, given Russia's late start and the fierce competition in the global renewables market from countries such as China. The conclusion seems well-nigh inexorable: the progress of renewables will be slow in Russia for the foreseeable future, exports will remain scarce, and the impact of renewables on Russian emissions of CO_2 will be insignificant.

The Worldwide Revolution in Renewables

Outside Russia there is a global revolution going on in renewable energy, chiefly in solar and wind for electrical power. It is on the same scale as the "shale gale" in oil and gas, discussed in the previous chapters, and it is far more widely distributed. It has been driven, above all, by advances in technology and know-how, and assisted by government policy in numerous countries. As a result, costs have plummeted. Over the decade of the 2010s, the costs of solar power declined by 85 percent, those of wind power by 50 percent, and battery technology by 85 percent, and they are continuing to drop rapidly.[11] Renewables (a category that includes hydropower) exceeded coal-fired power capacity as early as 2015, and the gap continues to widen. Gas is the next major competitor, but the ultimate outcome is already clear: in more than thirty countries, power from wind and solar, even without government subsidies, is already cheaper than power from gas.[12]

This has set off a tidal wave of investment in renewables worldwide. In 2019 alone, over $138 billion was invested in wind power, closely followed by solar at $131.1 billion.[13] All told, "clean energy" investment tripled in the last five years.[14] On the eve of the pandemic, renewables appeared unstoppable. According to the most optimistic forecasts, in a world that is steadily growing more electrical, renewables might account for as much as half of electricity output by 2050.[15]

The COVID-19 pandemic may slow the advance of renewables over the next decade, but it will not stop it. The progress of renewables is expected to take place in two stages. In the 2020s, natural gas, thanks to its abundance and low costs—in particular, cheap shale gas in the United States—will still have the advantage. But beyond that, in the 2030s and 2040s, renewables will gradually become dominant. The contest will increasingly depend on the rate at which renewable power penetrates sectors like home heating and industrial production. That process will be slower, as existing assets such as housing and factories are replaced or upgraded. But by mid-century, in a world that is rapidly becoming more electrified, renewables will prevail over gas.[16]

Renewables are not without their problems. The main one is so-called intermittency—the sharp drop in generation that occurs when the sun is not shining or the wind is not blowing. When that happens, utilities must scramble quickly to make up for the shortfall, typically with power from gas-fired plants, commonly called "gas peakers." Therefore, the main challenge ahead for renewables is storage—the capacity to store the excess power generated by wind and solar during their peak hours, making gas peakers unnecessary. This will require major advances in battery technology. The good news is that in this area as well, there is rapid progress and costs are steadily declining.[17]

The other major problem is transmission. To move electricity from regions of abundant wind and / or sunshine to areas of high demand, long-distance high-voltage power lines are necessary. These are expensive to build and increasingly unpopular with politicians and landowners, who demand compensation for the use of their land. Many years can go by while interconnections are delayed.[18]

Finally, there are economic limits to the penetration of renewables into the power balance. Renewables tend to self-cannibalize because they are generally clustered together in the same places. Moreover, over time, renewables must compete with the most efficient remaining fossil fuel power sources (the "residual fleet"), and their cost advantage disappears. By 2040 renewables will no longer be the cheapest source of power.[19] But by that time the battle will already have been largely won, with market penetrations in some countries of as much as 70 to 80 percent of electric capacity. Thus, from the standpoint of climate change, renewables loom as a game changer.

Except in Russia.

Renewables in Russia

In Russia the power sector is dominated by natural gas in the western third of the country, and by coal and nuclear in the eastern two-thirds. In the west, oil was largely displaced by natural gas in the 1980s, to free up additional oil for export. Since then, the share of natural gas in electricity generation has continued to grow, thanks in particular to the policy of providing gas to power plants at regulated low prices. Today gas fuels more than half of the power generation in Russia as a whole. In both east and west, the dominance of the existing fuel mix is reinforced by powerful political interests and established regulatory structures, as well as the enormous inertia of the existing infrastructure.[20]

Thus, the main near-term opportunity for renewables is in remote areas that are not connected to the central transmission system. This is a large land mass—about 60 percent of the country[21]—but it adds up to only about 10 percent of the population. Moreover, these are some of the poorest regions of the country, such as the North Caucasus and the Russian Far East. In the Far North, small communities generate power locally, primarily using subsidized diesel fuel, which is trucked in every season from the south (the so-called *severnye zavozy* that date back to Soviet times). Neither of these regions is attractive to private investors, and government support for renewables, though essential, is so far lacking.

One potential political asset for renewable technology, much touted by politicians, is that it could create a new market for Russian high-tech exports. Thus, renewables have been tied to Russia's "localization" policy, which aims to replace foreign imports with homemade products and skills, and promoters such as Chubais have used this potential as a selling point. But so far the result, as Chubais concedes, has only been to increase the number of bureaucracies involved and the complexity of regulations, and the consequence has been delays in actual construction.[22]

Most of Russia's efforts to date have gone into solar, but wind is starting to catch up. Since 2013 the capacity of Russian renewables has grown from 0.38 gigawatts (GW) in 2013 to 1.54 GW in 2019, a fourfold increase. (This may sound like a lot, but 1 GW is only enough to power a small town.) In 2019, solar and wind together produced 1.6 billion kilowatt-hours of Russia's total 1,080.6 billion kilowatt-hours, or 0.15 percent of the total.[23] On the eve of the pandemic, capacity had been projected to reach 5.4 GW by 2024, which would have been about 1 percent of Russian generation, although those targets will now probably be pushed back.

These are clearly very small numbers, especially when compared to Russia's much larger investments in gas-fired power. Thus, when Aleksandr Novak, who was then energy minister and is now deputy prime minister, met with Putin in early 2019 to report on the previous year's results, he contrasted Russia's modest 0.37 GW of solar and wind with the recent commissioning of two new nuclear power plants, for a total of 2.2 GW, as well as 2.2 GW of new thermal plants.[24]

What are the odds that a more powerful push for renewables could arise in Russia? That would require three things: more widespread investment and entrepreneurship, committed leadership support, and a strong economic case. There are elements of the first, but the other two are largely missing. Three further important drivers—public awareness, environmental urgency, and international diplomatic pressure—have until recently been almost entirely absent. Faced with the threat of carbon border taxes from the EU, the Russian energy sector has so far responded mainly with symbolic efforts—except for LUKOIL, which owns refineries in two EU member states, Bulgaria and Romania, and thus has a direct incentive to take decarbonization seriously.[25]

I focus here on the most positive part of the story—the rising entrepreneurship of a small number of individuals and companies, thanks to whom Russian renewables have achieved their first modest results to date.

A Handful of Visionaries

The story of renewables in Russia began in 2007. At that time Anatoly Chubais was the head of UES, which was the monopoly of the Russian power system, and in that capacity he masterminded the restructuring of the power industry, including the creation of a wholesale power market.[26] Chubais used this opportunity to insert a small space for renewables. The law on wholesale markets adopted in that year provided that, as he put it in a lecture a decade later, "the wholesale market could be a source for building a system of support for renewable energy." But as Chubais recounts the story, it took ten years of battles with the bureaucracy to translate that vague phrase into reality, and to coordinate the dozens of regulations and multiple ministries involved. "We had to appear five times before [president, and then prime minister, Dmitriy] Medvedev, when progress was being blocked. It was not until 2017 that we were able to win the battle. It was tedious, painful, and difficult bureaucratic labor."[27] The most hostile opponents, not surprisingly, were the establishment of the electricity sector. Chubais recalls, "They told our team, 'Why the hell are you meddling in our power sector with your stupid windmills, when it's clear that we need to devote all the money to our existing thermal generation, without any nonsense.'"

To make headway, Chubais and his team set up a variant on what is known as a capacity market. The first bid round for a renewables contract under this system was held in 2013. The winner—and only bidder—was the Italian company Enel Russia, which won two contracts for a wind project in Rostov Oblast (Rostov Province) and another in Murmansk.[28] But as late as 2017 there was still only one bidder. Then, in 2018, real competition appeared for the first time, which Rusnano won with the lowest bid. In 2019 the field expanded further, and there were eight times as many bidders as there were contracts available. This time Enel was again the winner, with a project to be located above the Arctic Circle, on the Kola Peninsula, and another in Stavropol Oblast in the south. Each year the winner won with a lower offered cost per kilowatt than the year before. In 2019 Enel won with an astonishingly low bid of only $65 per kilowatt. Even the state-owned nuclear giant Rosatom, which had never won a bid round, had lowered its bid each year.[29] Competition was evidently working.

The government is anxious to promote homemade technology; to achieve this, it has set up a system of sliding targets. The compulsory share of Russian-made components is designed as a ratchet system, which goes up year by year: it was set at 50 percent for solar power in 2015, then rose to 70 percent in 2016; for wind power, it was set at 25 percent for 2016, then rose stepwise to 65 percent in 2019.[30] Over time the government has increased its opposition to foreign investors, contractors, and suppliers in renewables. In the fall of 2019 the Ministry of Industry and Trade even circulated a proposal to zero out the share of foreign companies in renewables.[31]

With this system in place, the first renewables projects began to appear. The companies involved are of several different sorts, and they have a variety of motives. The dominant player to date is Chubais's recent company, Rusnano. Solar and wind have followed separate paths, so we begin with a look at solar, then at wind, before returning to the outlook for renewables as a whole.

Solar

An early supporter of renewables, mostly behind the scenes, was one of Russia's best-known oligarchs, Viktor Vekselberg. Having made a first fortune in the privatization of the Russian aluminum industry in the 1990s, Vekselberg was one of the main drivers in the rise of TNK (Tyumen Oil Company), which became one of Russia's largest oil companies. At the end of the 1990s Vekselberg consolidated his assets in energy, metals, and telecommunications under a single powerful conglomerate, Renova.[32] Since the beginning of the 2000s, Vekselberg has increasingly used his wealth and influence to promote technological innovation in Russia, notably by employing the resources of Renova to invest in a wide variety of high-tech ventures, which soon included renewables.

The beginnings of actual solar power in Russia go back to 2009, when Vekselberg formed a joint company, Hevel, with Chubais to build solar panels and ultimately solar arrays.[33] Vekselberg's Renova was the majority shareholder, with 51 percent, while Chubais's Rusnano took 49 percent. The initial capital investment was 7.5 billion rubles (then about $247.5 million), out of a planned total of 20.1 billion. Renova put in 2.8 billion rubles at the front end; Rusnano's contribution was a seven-year credit of 9.8 billion rubles at 13 percent yearly interest.[34]

The two partners joined forces for the construction of a factory and the purchase of materials for the production of thin-film silicon wafers. Hevel's plant was located in Novocheboksarsk, near the Volga River about 600 miles east of Moscow, on land leased from a subsidiary of Renova, Orgsintez Khimprom. Another business controlled by Renova, Oerlikon Solar, based in Switzerland, provided the thin silicon films and the turnkey facility used to build the first panels.[35] But from an early date Hevel also sponsored the development of Russian-made silicon wafers, with a grant to the Ioffe Institute in Saint Petersburg. This was apparently successful: the Russian-made solar panels are said to operate with a high 23 percent capacity factor (comparable to the worldwide average) and have won international recognition.[36]

Hevel began production in 2011, initially producing 120 megawatts (MW) worth of solar panels in that year. From the beginning the company was intended to produce Russian-made components for export as well as the domestic market. These were to be marketed by another Renova-controlled company, Avelar Energy Group, which sold renewables in southern Europe. Renova's vision was to build a complete value chain, from the supply of components by Oerlikon Solar to the marketing of Russian-made photovoltaics in Italy, Spain, and Greece through Avelar.[37] In early 2019 Rusnano sold off its stake in Hevel, in keeping with its strategy of investing seed money in a new venture and then helping it through its teething period until it is safely mature. Chubais was evidently satisfied that Hevel could now survive on its own.

Yet the way forward was bumpy. Beginning in about 2005, China, which had been a minor player in the solar industry until then, entered the global market.[38] Thanks to generous subsidies from the Chinese state, a host of fledgling Chinese manufacturers began producing solar panels. As domestic production boomed, they launched massive exports, at increasingly cut-rate prices. Between 2009 and 2013, at about the time Rusnano and Renova were launching Hevel, the prices of Chinese panels dropped by over two-thirds, upending the solar industry all over the world.[39] With solar energy suddenly a loser, Renova sold off its interest in Solar Oerlikon, and its subsidiary Avelar Energy exited from the solar business in southern Europe. Renova's plan to build a value chain for Russian solar, based on Renova's various subsidiaries and leading quickly to exports of Russian technology, abruptly collapsed. Today 80 percent of the world's solar panels are manufactured in Asia, mostly in China.

Nevertheless, inside Russia, Hevel has survived. It is the leading player in the Russian solar industry, and claims to have been the first to produce solar power.[40] By 2020 it had 611.5 MW of solar arrays in operation, and a further 1,150 MW of projects in its order book. An interactive map on its website shows thirty-five arrays operating in Russia, most of them in southern regions, roughly paralleling the northwestern border of Kazakhstan.[41] To put this story in perspective, again, these are still very small numbers. In 2019 the United States had 85 GW of solar capacity installed, nearly a hundred times more than Russia, and it was growing far faster.[42]

The other major player in Russian solar is a Chinese company, Solar Systems. Chinese companies have so far made little headway in Russian solar, partly because the economics are largely unattractive, and partly because the Russian government is wary of Chinese competition. Solar Systems has two solar arrays in operation, one in Stavropol and the other in Samara, and the company has approvals for a total of 365 MW in five southern oblasts.[43] To meet domestic-content requirements, the company built a 100 MW panel factory in Tatarstan, but it appears to source other components, such as inverters, from Chinese suppliers.[44] However, it also has a Russian partner, Solar Silicon Technologies, based in Podol'sk in Moscow Oblast. This is a rare example of a purely private Russian start-up: as Chubais described it, "a private investor took $100 million from his own pocket and built a factory," which produces silicon wafers for Solar Systems.[45]

Largely missing from the mix are residential and industry-scale renewables, especially solar. Rooftop solar arrays, which dominate the residential landscape in places like California and Hawaii, are almost totally absent in Russia. Most of the urban population lives in apartment buildings, which are poorly suited to rooftop arrays. The main market for residential solar would be suburban dachas, most of which are modest affairs. There is no domestic industry to supply solar systems to households, and most people cannot afford imported ones. Necessary incentives, such as feed-in tariffs, which allow home-owners to sell their home-generated power to the utility at a guaranteed price, have not yet been set up. According to Chubais, there is legislation pending before the Duma for so-called microgeneration, defined as less than 15 kilowatts, but there appears to be no sense of urgency about it.

There are limited efforts going on in isolated areas. Industry-scale renewables are starting to make an appearance, chiefly in the oil industry, which uses solar arrays to provide power to isolated oil fields. LUKOIL and Gazprom Neft have built industry-scale solar arrays at some of their refineries. Tomskneft has a wind generator and a hybrid system. Gazprom has installed small solar arrays to provide power to some of its gasfields at Yamburg, and to its pipelines in Stavropol and near Moscow. In the Far East, especially in Sakha, Rosgidro, the Russian hydropower monopoly, which has a powerful position in eastern Russia, is helping local communities build small solar arrays and wind parks, using equipment from Japan. All of these facilities are separate from the main grid and provide power for local consumption, replacing diesel fuel.[46]

In sum, Russia's efforts in solar power still consist mainly of isolated efforts. What has been lacking is a truly major player committed to solar. In wind, on the other hand, something different may be developing.

Wind

So far, wind has lagged far behind solar in Russia, but it shows signs of catching up. Today's wind turbines are giant structures, larger than long-distance airliners, and they involve high capital costs and advanced heavy-industrial engineering. For that reason, there have been no private investors or companies in Russian wind to date, except as suppliers.

The one potentially major player in the nascent Russian wind business is the state-owned nuclear giant Rosatom. Rosatom is one of the ten largest corporations in Russia. It is a vast state-owned conglomerate, employing more than 250,000 people in some 350 companies and organizations, and producing everything from water treatment to isotopes for medicine, and much more. But its core business remains nuclear power (as discussed in Chapter 7), and in recent years it has developed a thriving international business. In other words, Rosatom has plenty of experience in promoting exports. Rosatom also plays a lead role in the development of infrastructure along the Arctic seacoast. Logically, therefore, Rosatom might be expected to have an emerging interest in wind power in that region. Finally, Rosatom is a powerful engineering organization that has the industrial skills to build today's giant windmills. In Rosatom, then, wind power might ultimately find a strong champion.

However, that moment has not yet arrived; Rosatom's involvement in wind power is still at an early stage. It made a modest start in 2016 by establishing NovaWind, a subsidiary devoted to wind.[47] Under a contract with the Ministry of Industry and Trade (Minpromtorg), NovaWind has built a plant to produce components for wind turbines, drawing on the facilities of Atomenergomash, the nuclear engineering arm of Rosatom, located in Volgodonsk, in Rostov Oblast in south Russia.[48] Foreign partners are playing an important role in helping NovaWind tool up: for the final testing before its first start-up, NovaWind is using the company Red Wind, a joint venture of NovaWind and a Dutch company, Lagerwey. Rosatom is drawing heavily on foreign partners to develop its skills, sending its employees to Germany and Holland for training in windmill technology.[49] NovaWind is still a small operation, employing fewer than 300 people; its job is to consolidate the existing resources of Rosatom around wind, drawing on a wide range of Russian companies in the Rosatom empire.[50] For Rosatom, in other words, renewables are still a very small toe in the water.[51]

So far, most of Rosatom's effort has been concentrated entirely in south Russia. By early 2020 NovaWind was nearing completion of a wind park in the North Caucasus republic of Adygeya; at 150 MW it was the biggest in Russia. Adygeya is a good example of the kind of isolated location where renewables are the ideal solution: the region is chronically short of electricity, and the new wind park will add 20 percent to the republic's generating capacity. After completing the Adygeya wind park, NovaWind moved on to a larger project in nearby Stavropol, with a capacity of 210 MW. This is the first of four planned windmills, and will be Russia's largest wind farm.[52] Rosatom's total work plan is for 1 GW of capacity by 2023, for a total investment of 50 billion rubles (currently about $770 million).[53] This is serious engineering: the new windmills in Stavropol will be 150 meters high and weigh 400 tons. But Rosatom has plenty of experience with large projects.

Procuring Russian-made components is a top priority for Rosatom under its agreement with the Ministry of Industry and Trade. Rosatom is well placed for this, since it can call upon the resources of its vast organization. The Adygeya and Stavropol wind farms are advertised as consisting of 65 percent Russian parts, and are thus two of the few projects that meet the ministry's requirements.[54] Rosatom's goal is to reach 80–85 percent.

The real question ahead is whether Rosatom will be able to use its growing skills in wind to open up a new export market for Russia. So far there have been no reports of any projects outside Russia, or of exports of Russian-made components. As with Hevel in solar, NovaWind's first priority at this stage is localization—developing Russia's own capacity to provide the entire value chain, from components to operation, using wholly Russian-made components and skills.

The second major player in Russian wind is the Russian division of the Italian electricity giant Enel. When Enel Russia won the first bid round for renewables capacity held in Russia, it began building a 90 MW wind farm in Rostov Oblast, which so far is the company's only completed project. Enel Russia, based in Yekaterinburg, is the Russian subsidiary of the Italian electrical giant Enel, which owns 56 percent of it.[55] It has been active in Russia since 2004, but its projects until recently consisted of coal-fired and gas-fired power plants. As climate change has gained visibility, Enel has increasingly promoted an image as a green company. In Russia it sold off its coal-fired power plant in 2019 and has modernized its three gas-fired plants. But since 2017 its main activity in Russia has consisted of renewables. By the beginning of 2020 it had won tenders for three wind parks, for a total of 362 MW. The first, Azovskaya, located in Rostov Oblast, is about to begin production. The second will be located in Stavropol Province. Enel notes that its three wind parks will enable it to reduce its emissions by over a million tons of CO_2 annually.[56]

Rusnano, too, has played an active role in wind development, in addition to its pioneering activity in solar. Rusnano and the Finnish power company Fortum have formed a Fund for the Development of Wind Power, a 50–50 partnership of the two companies. It began with a small 35 MW wind park in Ul'ianovsk Oblast, which started up in 2018.[57] In Rostov Oblast, the Sulinskaia wind park is another joint project, which began operating in March 2020, using technology from the Danish company Vestas, the world's leading manufacturer of wind turbines.[58] Rusnano has since begun another wind project in the southern republic of Chuvashia. The Fund's latest activity is a wind park in Kalmykia.[59]

The parts for Rusnano's first wind turbine were purchased from a foreign company, Oerlikon in Switzerland.[60] But Rusnano makes localization a core priority of its wind program, and it has built three plants to supply Russian-made components for wind turbines; according to Chubais, they are now "producing in series." One of these, a plant in Taganrog in Rostov Oblast, began producing towers for wind turbines in 2018, using technology from Vestas and Siemens Gamesa.[61] Another plant, built by Vestas's Russian subsidiary Vestas Manufacturing Russ, employs 400 people in Ul'ianovsk Oblast, and built Russia's first composite turbine blades in December 2018. Rusnano is an investor in the plant, with a 24.5 percent stake in the 2 billion rubles investment, alongside Vestas and a "consortium of investors" from Ul'ianovsk Oblast.[62] The third plant, located in Nizhny Novgorod Oblast, produces gondolas (these are the parts of the wind turbine that control the windmill's direction and speed). These three homemade components will then be used for the next generation of the joint Fund's wind turbines. The ever-enthusiastic Chubais, at the 2019 annual meeting of Rusnano, proudly presented a photograph of Russia's first domestically produced wind turbine, 90 meters high with blades 60 meters long, on which he also briefed President Putin.[63]

The Ministry of Industry and Trade's moving scale of required domestic content creates a constantly rising benchmark for producers like Rosatom and Rusnano to increase the percentage of Russian-made parts. The first project built by the Rusnano-Fortum joint venture achieved 55 percent domestic content; its next project, the Sulinskaia wind farm, has reached 65 percent. But the ratchet will continue to go up in following years. Under the second renewables support program, Russia's wind projects must reach 90 percent domestic content by 2035, so manufacturers must keep running to stay up. Rusnano's latest initiative is yet another joint venture, this time with NordEnergoGrupp, a subsidiary of Severgroup, an investment company and conglomerate owned by billionaire oligarch Aleksey Mordashov. The new joint venture will invest in a plant to produce electrical equipment for wind turbines.

The cooperation of local officials is key for renewables—especially for wind, because of the large land areas required for the wind parks themselves as well as for the expansion of local power lines. Rostov Oblast is an emerging hub for Russian renewables, both solar and wind, thanks to its convenient location and the active support of the Rostov Oblast government. Rosatom's NovaWind has a partnership with the Rostov regional government to invest 2.9 billion rubles to develop a second facility to produce components for windmills, using technology from Lagerwey.[64] Rosatom and the Italian power company Enel also have projects there. The governor of Rostov Oblast, Vasily Golubev, is an experienced politician with a long prior career in Moscow Oblast, and is both well connected and supportive. It also helps that Rosatom is one of the largest employers in the oblast.

Although so far Russian wind power is located mainly in the south, the key to its future may lie in the north, and especially in the rapidly emerging field of offshore wind.

Offshore Wind: A Whole New Challenge for Rosatom

Offshore wind is the rising giant of the global renewables business. After a history of false starts, by the 2010s offshore wind had come into its own as the fastest-growing renewable in the world. By 2018 the installed capacity of offshore worldwide had reached 23 GW and was expanding at 30 percent per year,[65] and 5,500 offshore turbines were operating around the world. On the eve of the COVID-19 pandemic, according to the International Energy Agency (IEA), it was projected that by the 2040s total investment in offshore wind would top $1 trillion, rivaling gas-fired generation and coal. By that time, offshore wind would become the EU's largest source of electricity, and would have a strong position in China and the United States.

Long considered a costly curiosity, offshore wind has matured as a technology, with blades made of advanced composite materials and with complex electronic systems to adjust to changing wind speed and direction. Today's offshore wind turbines are some of the most formidable structures in the world, reaching as high as 220 meters, compared to 90 meters in 2010. By the 2040s the largest offshore turbines will be nearing the height of the Eiffel Tower.[66] Along with increased height and sweep have come dramatically lower unit costs.

So far Russia is absent from this scene. The main relevant source of expertise for offshore work is the oil and gas industry, but Russian oil and gas companies are only now beginning to develop the necessary offshore skills and experience, and they lack the homegrown industrial base to supply platforms and other equipment. Russian offshore work in oil and gas to date has relied largely on foreign contractors and providers, which lately has resulted in paralysis as a result of the US and EU sanctions imposed on any Western participation in the deep Arctic offshore. Another major obstacle is that offshore wind has much higher upfront capital costs than either solar or onshore wind; one 250 MW project costs about $1 billion—more than the entire Russian investment in renewables to date.[67]

Meanwhile, Russia's global competitors are off and running. The manufacturing segment is dominated by two companies, Denmark's Vestas (formerly DONG) and a Spanish-German partnership, Siemens Gamesa. Together these two players have over 80 percent of the world market. The field of developers and operators is more scattered, but the Danish company Ørsted is growing rapidly into a world leader.[68] The obvious strategy for Rosatom, if it makes a move into offshore wind, would be to partner with Vestas or Siemens, or with Ørsted, but at this moment there is no sign of any interest by Rosatom, which to date has focused solely on onshore projects.

Conclusions

The story of renewables in Russia to date is a case of low priority and many obstacles. Despite the efforts of Anatoly Chubais and a handful of promoters, renewables are still at their beginnings. In 2019 wind accounted for just 0.32 percent of Russia's electrical output, and solar for 1.3 percent.[69] Renewables have only a tiny part in the government's program for reducing CO_2 emissions. Even Chubais, despite his optimism, forecasts that by 2035, renewables will have only 12 GW of capacity, accounting for only 0.016 billion tons of CO_2 reductions, compared to the government's target of 0.545 billion tons for all reductions in the power fleet. The emphasis in the government's power policy lies entirely on improvements in efficiency (mainly from adoption of combined-cycle gas-fueled plants), not renewables.[70]

For the Kremlin and the Russian government, the major reason for promoting renewables is their potential to stimulate a new branch of high-tech industry for exports. According to Aleksey Khokhlov, who heads electricity research at the Skolkovo School of Management, this localization policy takes precedence over the addition of generating capacity, for which there is little real need.[71] In practice, the government's policy has been more of an obstacle to investment than a stimulus.

From the standpoint of climate change, the minor role of renewables has two consequences. The first is that for the foreseeable future, renewables will make only an insignificant contribution to limiting CO_2 emissions from the Russian power sector. The second is that Russia is still very far away from being able to export renewables technology to the world market and thus earn export revenue. To keep it growing will require continued strong support from the state, otherwise, as Chubais says, "the baby will suffocate." In short, Russian renewables are not yet viable on their own. As for Anatoly Chubais himself, now that he has left Rusnano, only time will tell whether he remains a major player in Russia's climate-change politics, or whether his remarkable career is now finally nearing an end.

6

THE REVIVAL OF RUSSIAN NUCLEAR POWER

On August 17, 1998, the newly named prime minister of Russia, a young thirty-five-year-old named Sergei Kirienko, stepped up hesitantly to face the massed microphones and cameras of a crowded press conference, and announced that the Russian government was declaring a moratorium on its domestic debts and would no longer defend the ruble. In effect, he was declaring bankruptcy. The Russian government was deeply in debt, having financed its budget with a growing mountain of short-term government bonds. On that day, the whole tottering structure came crashing down.[1] One week later, Kirienko was out of a job. He had lasted exactly five months.

Following his dismissal, Kirienko's prospects did not look good. When Vladimir Putin—another Yeltsin surprise—was elected president in 2000, he dispatched Kirienko back to his native Nizhny Novgorod as presidential envoy to the newly formed Povol'zhe Federal District, a form of honorary exile. Kirienko's Moscow career appeared to be over.

But in 2005 came a surprising turnaround: Putin unexpectedly named Kirienko to head the Russian agency for nuclear power, Rosatom. Kirienko had no more background in nuclear power than he had previously had in government,[2] apart from a long-past first job as a young man at a submarine-building plant.[3] But as Putin said at the time he appointed him, "The point is not to turn Kirienko into a nuclear guy [*atomshchik*]. . . . It's all about the organization of the industry . . . and Kirienko is the man who can handle that task."[4]

Kirienko's appointment proved to be an inspired move. He served as head of Rosatom for eleven years, from 2005 to 2016, and was the chief architect of a remarkable renaissance of Russia's nuclear power industry. That renaissance, the rise of Russia's nuclear champion Rosatom under Kirienko's leadership, and the resulting dominant position of Russian nuclear technology in today's international market, are the subjects of this chapter.

This story might appear at first sight to be far removed from the main subjects of this book—climate change, energy transition, and Russia's prospects in tomorrow's world—were it not for the fact that by mid-century an increasingly climate-stressed world might do what seems unimaginable today—turn to nuclear power as its last hope to avoid a wholesale climate catastrophe. If that day comes, Russia could find itself in a leading position, and Russian nuclear power could be one of its few opportunities for a major high-tech export. Not only would this help to offset the decline of Russia's revenues from fossil fuel exports, it would also enable Russia to make a contribution—one of its few—toward a decarbonized world.

That would be, to say the least, a major positive result in a high-tech story that so far has not seen much good news. Russian nuclear power stands as virtually the sole exception to Russia's otherwise weak place in the development and export of world-class technology and innovation. To most readers, for whom the memory of Chernobyl is still fresh, the rebirth of Russian nuclear power will seem all the more surprising. To understand how this happened requires some explanation, to put the Russian nuclear story in its domestic context. But first we need to take a look at the global picture.

The Global Context of Nuclear: Decline in the West, Boom in the East

Nuclear power is the second-largest source of low-carbon electricity in the world today, with 452 operating reactors generating 2,700 terawatt-hours of electricity in 2018, or 10 percent of global electricity supply.[5] It makes an important contribution to lowering the world's carbon footprint, and therefore logically should play a large part in any decarbonization strategy. Some observers argue that nuclear power should be the centerpiece. These include the nuclear industry, naturally enough, but also a handful of others who believe deeply in the threat of climate change but are skeptical of the world's current approach to dealing with it, and see nuclear power as a leading part of the solution

Bill Gates—the founder of Microsoft—is the best-known example. As a frequent commentator on clean energy, he has campaigned for over a decade in favor of a "nuclear first" strategy. His reasoning is that "zero net carbon" policies impose unreasonable costs on the world's population and will face increasing political resistance; in addition, they are beyond the means of developing countries. Consequently, they will fail. Gates predicts that by 2050, instead of the 1.5 to 2 degrees Celsius warming called for by the 2015 Paris Accord, the world will face an increase of 5 to 7 degrees, which will only worsen in decades to come. The only sensible answer, Gates believes, is nuclear power.[6]

Another convert to nuclear power is Anatoly Chubais, whom we met in Chapter 5 as Russia's chief advocate for renewables. As Chubais told an audience in a 2019 lecture at the Bauman Moscow State Technical University, "In the 1980s, I thought nuclear was over, and I've spent my life 'competing' with nuclear, but I've changed my mind."[7]

Nuclear power has a number of familiar advantages. It provides a steady supply of "baseload" power—the part of the electricity supply that is "always on"—regardless of the season or the time of day. It is arguably safe to operate: after all, despite its reputation, there have been only three major nuclear accidents in the past half-century, and the next generation of nuclear power plants promises to be safer still. As already noted, the operating costs of nuclear power are lower than those of any competitor source of power. Nuclear power does not depend on imports from an unstable Middle East or other geopolitically exposed places. And above all, it produces no greenhouse gases. For its advocates, nuclear power is the ultimate renewable; its time, they argue, has yet to come.

Gates's view, however, is not widely shared, at least in the West. Nuclear power has been out of fashion for over two decades, because of its high construction costs and long delays as well as lingering doubts over its safety and the unresolved problem of storing nuclear waste. Even though it represents 18 percent of electricity supply in the so-called advanced economies,[8] its share is expected to decline. In the United States and the European Union, where most of today's nuclear power is located, investment in nuclear is at a virtual halt. The average nuclear plant in the advanced economies is over thirty-five years old, and as much as a quarter of the existing nuclear capacity could be shut down by 2025.[9]

Meanwhile, there is virtually no new construction going on. In the United States, no new nuclear power plants have been commissioned in the past three decades. The only currently active project in the US, the Vogtle 3 and 4 project in Georgia, has become an embarrassing white elephant, sustained only by massive propping up by government loan guarantees. The project has been under construction for fifteen years; it was initially expected to cost $9.5 billion, but as of 2018 the estimated price tag had trebled to $27 billion, with no end in sight.[10] An earlier project under construction in South Carolina was abandoned in 2017.[11] Similarly, Hinkley Point C in the United Kingdom[12] and Flamanville in France,[13] both based on unproven new generations of technology, have been plagued by delays and cost overruns.

Only massive investment could prevent what the International Energy Agency (IEA) calls a nuclear fade scenario in the advanced economies (i.e., a steady decline in the remaining share of nuclear power), yet the IEA does not foresee a massive investment in nuclear forthcoming in coming decades.[14] For the IEA and much of the energy community in the West, nuclear power has no significant role to play in the developed world's response to climate change, although the IEA warns that a nuclear fade scenario would add major costs to the decarbonization program.

The developing world is another story. Nuclear power capacity will more than double in China and twenty other countries by 2040, predicts the IEA, more than offsetting the decline in the advanced economies.[15] Even in the developing world, however, nuclear power faces a number of question marks. Renewables and even coal will be major competitors, and the high costs and frequent delays of nuclear construction present serious obstacles, raising questions about whether the necessary investment will be available.[16] Technological innovations, such as small modular reactors, may open new, lower-cost opportunities, but these are still on the horizon.[17] At best, concludes the IEA, the share of nuclear power will be only 5 percent of total electricity capacity in the developing world by 2040, even under a sustainable development scenario.

Overall, the consensus among official agencies and consultancies in the West is that by 2040 nuclear power will supply at best the same share of global electricity as today—between 10 and 12 percent.[18] However, by that time total electricity generation will have increased by half compared to today, to between 37,000 and 39,000 terawatt-hours. The implication is that, just to maintain its current share, nuclear generation would need to increase by 50 percent as well, as mentioned above, doubling in the developing world to offset the decline in the advanced economies.[19]

The implications of this twofold picture for Russia are far-reaching. First, insofar as there is a growth market for nuclear power, it will be in the developing world, above all in China. Second, the major challenge will be cost, as nuclear competes with renewables and coal. To achieve lower costs, cheaper and more efficient designs will be at a premium, possibly centered on small modular reactors. Third, a major constraint will be the limited availability of financing, which will make large-scale support from the Russian government a necessity. Finally, the impact of climate change will be decisive in determining the world's attitude toward nuclear. The only strategy for Russia, then, is the long game—to focus on the developing world, while successfully managing the challenges of cost and safety—and positioning itself for the longer-range prospect of a global renaissance of nuclear power. By mid-century, if this scenario is realized, nuclear power in Russia could be a badly needed source of export revenue.

In contrast, there is likely to be little demand for any significant expansion of nuclear power inside Russia. The overall growth of electricity demand is likely to be slow, and the abundance of cheap gas will keep nuclear uncompetitive. (I explore this point in more detail below.)

Rosatom's strategists are fully conscious of this picture.[20] But will the Russian nuclear power industry be able to live up to these challenges? In the rest of this chapter we look at the remarkable turnaround of the Russian nuclear sector over the last thirty years, and its prospects for the future. I will begin, however, with an explanation of why nuclear is the exception that proves the rule.

Russia, the "Resource Appendage"

Russia has one of the world's most advanced scientific and engineering cultures. Its people are highly educated; its scientific institutions have traditionally been well regarded; and it has distinguished experts in many fields, who have made many important discoveries and inventions. But throughout its history Russia has typically lagged in turning its discoveries into commercial applications.[21] Moreover, its scientific and technological institutions were seriously weakened by the Soviet collapse, from which they have only partially recovered even now, thirty years on. Russia generates few major advances, and it is not a leader in any of the leading-edge technologies that are changing the world.

This can be clearly seen in its pattern of foreign trade. Russia's position in the global economy is largely that of a resource provider. Its exports consist mainly of raw materials—oil and gas and metals, basic chemicals and fertilizer, and lately wheat and corn—while it imports much of its technology. In manufactured goods and machinery and advanced chemicals, Russia's imports outweigh its exports by two-to-one, and in the more advanced categories, such as computers and software—the stuff of the internet and social media—practically everything is imported.[22]

What is even more striking is the small size of Russia's trade in technology in absolute terms. Its trade turnover in this category was only $3.8 billion in 2016 (the most recent year for which figures are available), compared to $220 billion for the United States, putting Russia between Portugal and Greece.[23] Russia remains largely closed off from world technology trade. This pattern has not changed much since Soviet times, and indeed was characteristic of pre-revolutionary Russia as well.[24] Russian critics have long had a name for Russia's position in the global division of labor: Russia, they say, is a *resursnyi pridatok,* a "resource appendage." The Russian foreign-trade signature places Russia squarely among the commodity-dependent countries of the world.

Russia's weakness in advanced exports is matched by anemic innovation throughout much of the Russian economy. This is due partly to weak commercialization of inventions by the R&D system and partly to Russian industry's resistance to adopting homegrown technology.[25] For Russian reformers, much of the fault lies with Russian elites' addiction to the easy profits from raw materials, especially oil and gas. It takes no brains, they say, to drill a hole in the ground. In a sense the charge is unfair: Russia's oil and gas industry, as we saw in Chapters 2 and 3, is highly technological. Russia's *neftianiki* and *gazoviki* (oil and gas workers) have successfully mastered many advanced techniques over the last two decades—but very largely using imported equipment, technology, and contractors. The same pattern can be seen throughout the economy, and especially in high-tech manufacturing.

Importing foreign technology does not necessarily spell backwardness. On the contrary, in the global division of labor, high-tech producers and exporters throughout the world are themselves dependent on imports of technology from subcontractors for much of their value chain. This gives subcontractors an opportunity to innovate themselves by copying imported technology. The outstanding example is China: Chinese companies copy foreign technology and leapfrog ahead, bypassing the original and generating world-class competitors in a short time. Dependence on foreign imports does not necessarily imply stagnation, provided it leads to innovation that produces a homegrown leader, combined with successful commercialization of the results. Yet there are few examples of that in Russia; instead, Russian manufacturers and producers continue to depend on imports without innovating from them.[26] Faced with sanctions, Russia is increasingly turning to newly developed Chinese technologies.

There is one major exception to this pattern—Russian nuclear technology, and especially civilian nuclear power. Not only has the Russian nuclear power industry been an innovator at home, but it has developed a thriving export business. What explains its success?

Russian Nuclear Power: Decline and Revival

The Russian nuclear power industry was initially born as a by-product of the Soviet military program. The first large Soviet nuclear power station began operation in Siberia in 1958; its precise location was secret because its primary purpose was the production of plutonium for nuclear bombs.[27] The close association of civilian nuclear power with the military had two contrary effects. Civilian nuclear power benefited from the overall high priority of the military-industrial complex in obtaining equipment and materials, but this came at a price: the nuclear empire was shrouded in secrecy; and military orders came first, while efficiency and safety took second place.[28]

In 1986, the year of the Chernobyl disaster, the Soviet Union had 30 gigawatts (GW) of nuclear capacity in operation, about 9 percent of the country's total power capacity,[29] compared to more than 83 GW of nuclear power in the United States (representing over 15 percent of US capacity in the same year).[30] The main objective of the civilian nuclear program at that time was to provide power to locations that were poor in fossil fuels, mainly in the western USSR (including some countries, such as Lithuania and Ukraine, that are independent today). The industry had two basic models—the ill-fated, graphite-cooled

RBMK that was used at Chernobyl, and a pressurized-water reactor called the VVER, which in various modern versions is still the anchor of the Russian nuclear industry today.

The end of the Soviet Union in late 1991 was a disaster for Russia's nuclear industry. The nuclear community, which consisted of hundreds of institutes and power plants and employed over a million people, many of them working in secret cities known only under code names, abruptly came to a halt, leaving their workers stranded. The previously all-powerful USSR Ministry of Nuclear Engineering and Industry, renamed Minatom in 1992, was ignored by a government that, as we have seen, teetered on the edge of bankruptcy and had more pressing concerns. During this period the nuclear industry supported itself mainly through its still-modest exports (part of which consisted of the sale of enriched uranium from dismantled nuclear warheads, exported to the United States). These brought in a welcome dollar flow of nearly $3 billion per year, but part of that revenue was commandeered by the Ministry of Finance to support other needs, including Russia's costly war in the Caucasus. At the end of the 1990s, the minister of atomic energy, Evgeny Adamov, estimated that the nuclear sector was receiving less than half of the funding it needed to support basic operations, not to mention additional tasks that had been assigned to it, such as dismantling nuclear submarines.[31]

As a result of the general disorder, the nuclear sector began to fray. There was a massive loss of skills, as experienced workers left for jobs in the civilian sector. Institutes and factories themselves took on non-nuclear business. Some key factories were taken over by private-sector interests, although the core of the industry remained intact under state control.[32] When Kirienko was parachuted into the top job at Rosatom in 2005, he found an industry that had lain dormant for two decades and had partially disintegrated. The core of it was still controlled by the remnants of the Soviet-era "Ministry of Medium Machine-Building" (Minsredmash)—the euphemism used to describe the body that controlled all nuclear matters, both military and civilian. But key assets had been taken over by private entrepreneurs and had been turned to other purposes.[33] Still others had fallen under the control of oligarchs. One of them, a picturesque Russian-Georgian former biologist named Kakha Bendukidze, had put together a conglomerate called United Machine-Building Factories (OMZ) that included the famous nuclear-engineering complex called Izhorskie Zavody. All told, the nuclear industry had completed only five new domestic units in the previous two decades, and much of that involved finishing work that had been undertaken in the 1980s before the Soviet collapse.

On the eve of Kirienko's takeover, Rosatom's international business was stagnating as well. Its three major legacy projects, in China, India, and Iran, were severely behind schedule, and as a result its international earnings were stuck at about $3 billion per year, much of that, as mentioned above, derived from reprocessing nuclear fuel.[34] The key player in Russia's foreign operations, Atomstroyexport, had been taken over by a consortium led by Bendukidze, and industry veterans complained that Atomstroyexport was being plundered.[35] This event evidently touched a nerve within the Russian government; one analyst calls the acquisition of Atomstroyexport a "fateful misstep" for Bendukidze. It was a wake-up call for Putin, and led directly to his plan to relaunch the nuclear industry, and his appointment of Kirienko to carry it out.[36]

What Kirienko Did at Rosatom

Kirienko's first task was to regather the nuclear industry under one roof. Bendukidze was pressured to sell OMZ to Gazprombank (which acted as an agent for the transaction), and his OMZ conglomerate was dismantled, with the nuclear portions turned over to Rosatom. At about the same time, Rosatom gained control of Atomstroyexport, which handled nuclear exports.

Next, Kirienko needed to bring order to the industry's supply chain. He faced a collection of about 80,000 equipment manufacturers and service suppliers that were monopolies in their fields and had a stranglehold on Rosatom. In 2007–2008, according to Kirienko, 85 percent of the nuclear sector was controlled by these monopoly providers. They charged high prices and set production schedules as they pleased. As late as 2010, half of Rosatom's supply contracts were behind schedule. Kirienko fought back by creating alternative engineering and equipment contractors under Rosatom, by buying up some of the providers, and by conducting tenders to stimulate competition. This brought fierce opposition from the incumbents, but by 2010 Kirienko was able to declare, "There are no more monopolies left in the industry." By the time Kirienko left in 2016, much of Rosatom's huge supply chain had been brought under a single system of competitive tenders.

Kirienko's most sensitive task was to disentangle Rosatom's civilian business from its military activities. When Rosatom was created in 2007 by a presidential decree, it inherited the entirety of the country's nuclear weapons assets.[37] In a policy called "The New Look," Rosatom undertook to separate the military wing from the civilian. Both would remain under Rosatom, but while the military part would continue to receive direct government funding, the civilian part was supposed to become self-supporting.[38] Yet the separation was easier to achieve on paper than on the ground. Many of the components of Russia's nuclear supply chain still served both civilian and military customers inside Russia, beginning with the mining of uranium, which was still largely handled in closed cities, for which Rosatom was also responsible. At the time of Kirienko's appointment, many industry veterans were skeptical that such a separation was even possible. Viktor Mikhailov, the former minister of nuclear power, told an interviewer that in his opinion the civilian and military halves were "inseparable" and that Kirienko would find the task "very difficult."[39] Another pointed out that the civilian business had originated as a "side-product" of the military program, and all the civilian technology had come from the military. In effect, this view ran, the two were connected at the hip.

Separating the weapons program from the civilian nuclear power was crucial, if Rosatom was to develop its international business on a commercial basis. Russia needed the endorsement of the international community, especially that of the International Atomic Energy Agency

(IAEA), which Russia had joined in 1957.[40] This in turn required meeting international standards for transparency and safety—the latter being especially important, in view of the lingering international memory of Chernobyl. Today Rosatom's military side, following Kirienko's restructuring, is largely kept administratively separate from the civilian, notably through the use of contracts, through which the military side supplies a wide range of civilian needs, including those of Rosatom's own civilian programs.[41] Rosatom, in return, provides essential services for the military, such as the decontamination of 200 retired nuclear submarines. Rosatom officials are discrete about the agency's continuing military roles, but it is noteworthy that when Kirienko reported to Putin, as he did regularly every year, the military contracts—the so-called *gosoboronzakaz*—were always the first thing Putin asked about.[42]

This effort also required a major restructuring of the civilian business, to put it on a financially self-sustaining basis. The industry's traditional culture had been all about engineering and had never paid attention to costs. Kirienko commented diplomatically, shortly after taking over Rosatom, "The sector's level of patriotism brings admiration; and its professionalism—high respect; but its level of understanding of the economy—quiet horror."[43] To drive the industry to change its culture, Kirienko created a special commission on restructuring, headed by a team of outsiders, who built a system of cost controls by which industry managers would be judged.[44]

Rosatom Looks Abroad

Kirienko's final task was to redirect Rosatom's priorities toward its international business. This had not been part of his initial plan. On the contrary, his first priority had been domestic nuclear power. He began with high ambitions. In early 2006, a few months after his appointment to Rosatom, he submitted to president Putin a vast program to relaunch domestic nuclear construction; he vowed to build forty new reactors in Russia by 2030, raising the share of nuclear in Russia's electricity output from 16 to 25 percent.[45] In June 2006 Putin approved the program. During the first phase, Russia would complete ten new 1,000 MW reactors by 2015, with ten more to begin construction over the same period. The program would cost $55 billion, nearly half financed by direct state subsidies and the rest from increased tariffs to consumers and private financing.[46]

But the expected demand for new domestic nuclear capacity failed to materialize. Beginning in 2006, growth in overall power demand began to slow.[47] By the end of that year, the domestic nuclear program had to be scaled back, and Kirienko, who earlier in the year had touted the prospects for the domestic market, realized that it was going nowhere.[48] But there was worse news ahead: the global financial crisis of 2008–2009 suddenly brought a halt to Russia's oil-fueled economic boom, and overnight there was practically no need for new power capacity at all. For a time Kirienko clung to his domestic strategy. As late as 2009 he vowed that the company would still build one new reactor per year. But in the end, even this more modest goal could not be met.[49]

At that point Kirienko began redirecting the company's priority toward the international market. The outlook for Rosatom's overseas business did not look promising. Atomstroyexport had been losing money since 2007, mainly as a result of construction contracts concluded in the 1990s with Iran, India, and China. In 2012 the company reported losses of 13.8 billion rubles (then about $500 million).[50] But by 2013 the situation began to turn around, partly as the result of a series of reorganizations, which by 2016 led to the creation of a new Engineering Division, an integrated unit that included design, sales, and service functions for all of Rosatom's overseas operations.[51]

Today Rosatom's overseas expansion is regarded as Kirienko's most signal achievement. Nuclear power has been a rare success story for Russian high-tech exports over the past decade. Back in the late 1980s, in the aftermath of the Chernobyl disaster, the reputation of Soviet nuclear power appeared shattered forever. Yet thirty years later, Russia has established itself as the global leader, much to the dismay of would-be competitors, particularly France and the United States.[52] At the end of 2019, on the eve of the COVID-19 pandemic, Rosatom had 34 GW of nuclear capacity in operation outside its borders. It boasted thirty-six units under construction in twelve countries, with an order book totaling over $130 billion and annual revenues of over $6.5 billion per year.[53]

Rosatom's international business was initially concentrated in the countries of the former Soviet Union, notably Ukraine, where 40 percent of Rosatom's international capacity is still located. But in the last decade Rosatom has expanded its range of customers to include not only Russia's near neighbors, such as Hungary and Finland, but also developing countries. At the beginning of the 2020s Rosatom had projects under construction in Turkey, Bangladesh, and Egypt, as well as new projects for long-established clients such as China and India.[54]

To the countries it courts, Rosatom offers an attractive "full service" package. In addition to building the power plant, it provides funding, trains local personnel, supplies the fuel, and handles reprocessing and waste treatment by repatriating used fuel rods back to Russia.[55] Unlike its Western competitors, Rosatom is able to deliver completed projects on time and under budget, thanks to a "cookie cutter" strategy that uses Russia's proven boiling-water (VVER) technology, which Rosatom has continued to improve over the years, rolling out successive generations of VVERs with improved efficiency and safety features. At a time when developing countries are increasingly aware of the threats of climate change and pollution, many are finding the Russian nuclear offering appealing. It is, of course, a growing foreign-policy asset for the Russian government, yet the Russian comparative advantage in nuclear power is above all economic and commercial.

Rosatom's excellent international safety record to date has enabled it to overcome the memory of Chernobyl and to weather the shock of the 2011 disaster at Fukushima, which caused a backlash against nuclear around the world. In 2012, in a television interview, Kirienko expressed optimism about the global prospects for nuclear power.[56] Rosatom's own international contracts, he claimed, had doubled in the previous year. The outlook was bright: Just replacing the nuclear power plants in operation around the world, he said, would require 320 to 350 new units by 2030 to keep the share of nuclear at its present level. Conscious that after Fukushima, safety and environment would be uppermost in the minds of clients, Kirienko announced proudly, "We have installed automatic systems for radiation monitoring at all of our reactors, and the data is automatically transferred to the IAEA in real time and on the internet. Anyone can access the internet right now and look up the current radiation readings."

But for Rosatom to continue expanding its international business, it must meet a number of challenges. The first is costs. Kirienko told president Dmitry Medvedev in a meeting of the latter's Innovation Commission in 2009, "Our main comparative advantage is our ultra-low operating and fuel costs; only solar is lower. But our main competitive liability is high capital costs; therefore our top priority is to lower our construction and start-up costs." The answer, Kirienko, said, is computerization of every phase of nuclear power construction, from three-dimensional design and construction to fully computerized management of supply chains, finance, and staffing.[57] Consequently, over the subsequent decade, digitalization has been one of Rosatom's top priorities.

The next challenge is innovation. Until now, Rosatom's development program has been incremental, consisting essentially of upgrading successive generations of its VVER pressurized-water technology. Russia's workhorse today remains the VVER 3+, developed under Kirienko. But soon it will need to move on to the next generation of nuclear reactors and fuels. Under Kirienko Rosatom also began modernizing its fast-neutron (BN-800) reactor technology, and by 2020 its first BN-800 was scheduled to begin operation in Sverdlovsk Oblast. According to Kirienko's successor as Rosatom's general director, Aleksei Likhachev, the BN-800 will soon be the company's standard offering for overseas projects.

The ultimate test of Rosatom's ability to meet these challenges and to continue dominating the international market will be competition from China.[58] Over the past twenty years, China has rapidly turned into a major force in nuclear power itself. China presently has nearly 49 GW of nuclear capacity in operation, up from only 2 GW in 2000, putting it in third place in the world. It is currently building eight new reactors per year, and is shooting for 120 GW by 2030. Currently, 85 percent of its capacity comes from domestic companies, compared with only 1 percent when it started out in 1996.

Russia built two initial units in China (Tianwan 1 and 2), which began operation in 2007, at a cost of $3.2 billion, with China contributing $1.8 billion. Two further units at Tianwan (Tianwan 3 and 4), with Russia contributing about 30 percent of the equipment, began operation in 2017 and 2018. Russia remains competitive, with a 2016 agreement to supply VVER-1200 units for Tianwan 7 and 8, although construction has not yet begun. But at this moment there is no further business lined up. A 2009 agreement with Atomstroyexport to build a fast-neutron reactor in China based on Russia's BN-800 technology did not go forward. An initial 2014 agreement to build floating nuclear cogeneration plants based on Russian technology and Russian-supplied fuels also did not proceed, when the Chinese decided to use their own designs.

Meanwhile, China is developing its own export capability, in direct competition with Russia, and using Chinese technology derived from French and American models, not Russian. For the time being, the only actual Chinese project in operation is in Pakistan, but negotiations are under way with eleven other countries, including Romania and Kazakhstan, which would previously have been considered part of Rosatom's business domain. China's export plans are constrained by the fact that China is not yet a member of the IAEA Vienna Convention on Civil Liability for Nuclear Damage (to which Russia is a party), and China is not yet capable of accepting used fuel (as Russia does). But these problems will no doubt be resolved over time, and China's readiness to supply generous financing will present a formidable challenge to Russia's established export position.

But the competition between Russia and China in the nuclear field will increasingly be determined by the world's responses to energy transition and climate change. We turn to those now.

Nuclear Power and Climate Change

Until recently, Rosatom's senior managers—Kirienko's alumni—hardly thought of themselves as apostles for sustainable development, but like other senior managers of Russian companies with international exposure, they have recently begun to emphasize their green credentials. In 2020 Rosatom issued its first sustainability report focused on ESG (environmental, social, and corporate governance). The report stresses Rosatom's role as part of the world's defense against climate change. Rosatom's deputy chief of strategy, Roman Golovin (who, like a growing number of young experts in the strategic departments of Russian energy companies, has worked in Western companies and graduated from Russia's top schools)[59] puts strong emphasis on the danger from CO_2 emissions and climate change. Golovin's public vision for nuclear power—like that of Western consultancies such as Bloomberg New Energy Finance (BNEF)—is that the future consists of renewables and nuclear. While Golovin concedes that renewables are increasingly competitive with fossil fuel generation, he argues that renewables alone will not suffice—they take up too much space, and they suffer from the problem of intermittency. (In this respect he is in accord with Putin's own expressed views, as we saw in Chapter 1.) Consequently, the future must be increasingly nuclear, and for Golovin, it will be Russian.

But what if this vision is not realized? Inside Russia itself, as we have seen, nuclear power has limited growth potential. Outside Russia, Rosatom will face increasingly powerful competitors. In response, the company is making efforts to diversify its own domestic activities beyond nuclear power. Rosatom's official corporate goal is that 30 percent of its revenue by 2030 will come from non-nuclear businesses.[60] In earlier chapters we saw Rosatom's growing involvement in the development of the Northern Sea Route (which includes nuclear icebreakers but also new ports and channels in the Far North), as well as wind, through its NovaWind subsidiary. Recently Rosatom has announced the creation of a new subsidiary under TVEL called Renera, which will manufacture batteries for electric vehicles, as well as storage batteries to provide peaking power for renewables.[61] Rosatom is also experimenting with hydrogen, to be produced by nuclear power in the Far East and exported to Japan.[62] These new entries will accelerate the growth of Russia's still-embryonic renewables business, but also potentially create new products for export.

The company's top priority outside the nuclear field is nanotechnology, consisting of laser technology, precision machining, and powder metallurgy, all of which are potential areas of strength for Rosatom. The most promising global application of nanotechnology is 3-D printing, which is revolutionizing manufacturing the world over. Rosatom has created a new subsidiary, Rusatom-Additive Technologies (RusAT), which heads a pool of some twenty research institutes, bringing together all of Rosatom's relevant skills under one roof. A single factory, Tsentrotekh, in Novoural'sk (Sverdlovsk Oblast), has been chosen as the production site for Russia's first large-scale 3-D printers, using laser additive technology.[63] RusAT has been placed under a senior figure in his sixties, Aleksei Dub, a State Prize laureate in the field of metallurgy, who at the same time is responsible for much of Rosatom's overall technology policy. There is no mention of "start-ups."

RusAt is running into some familiar problems. One is consumer resistance to 3-D printing as a new technology. "Producers are unwilling to take risks by departing from familiar production methods," says Dub. "Unfortunately, even though we organize displays and develop publicity, our information doesn't reach the companies. For many of those we do manage to reach, it comes as a surprise that additive technologies are no longer a fantasy."[64]

The second problem is that RusAT is under growing pressure to substitute Russian-made technology for foreign imports. For the time being it is sourcing most of its components "from the open market" (by implication, from abroad), and its lasers are purchased from IPG Photonics, a US-based multinational, although RusAT vows to displace these as soon as possible.[65]

Floating Nukes: A Solution without a Mission?

The idea of building floating nuclear power plants has a long history going back to Soviet times, but like the rest of the nuclear sector it went dormant after the collapse of the Soviet system. The logic of the concept is that a floating power plant would enable the development of the Russian north by providing power to ports, factories, and military bases along the Arctic coast.

The concept was revived when Vladimir Putin, shortly after his initial election in 2000, visited the Sevmash shipyard, the leading military shipbuilder in Soviet times. Putin figured that floating nuclear plants fit squarely with his overall vision for the Far North, and he has returned repeatedly to this theme. Rosatom responded with a plan that envisioned twenty floating power plants deployed throughout the Far North and the Far East.[66] Construction of the first floating plant, designed simultaneously as an icebreaker, began in 2007, at Sevmash. As originally conceived, the ship, named the *Akademik Lomonosov,* was intended to supply power to the western city of Severodvinsk and Sevmash itself. But it ran into problems almost immediately. The design was new, and Sevmash had little experience in civilian work. By 2008 little progress had been made. Rosatomflot broke off the contract and transferred the project to the Baltic Shipyard. Yet even then the path was not smooth, and it took another twelve years for the *Akademik Lomonosov* to be completed.[67] By that time additional power for the Severodvinsk site was no longer needed, and Rosatom had to cast about for other locations for the floating power plant to operate, eventually settling on the remote port of Pevek, in the Chukchi Sea. In January 2020, amid great fanfare, power from the first floating plant began to flow to Pevek and the surrounding region, where it will ultimately replace the aging Bilibino nuclear station.[68]

But the floating nuclear program looks like a solution without a mission. After much study Rosatomflot identified a half dozen locations along the Arctic coast that would be suitable, but there are currently no plans to build any more floating vessels. A trial concept to develop a smaller model that could be used in Siberian rivers has gone nowhere. Another concept, of potentially greater interest to Rosatom as a possible source of exports, is to use floating nuclear stations to supply power to offshore oil platforms. This idea was included in a package of agreements signed in May 2014 at the time of Putin's visit to China. Rosatom followed up with an agreement with the Chinese company CNNC New Energy[69] to form a joint venture to build six floating power plants. But the Chinese reportedly insisted that the agreement include a transfer of the technology to China, and the Russians backed away. There have been no further reports on the proposed joint venture since then.[70]

The slow progress of the floating nuclear program, compared with the much higher level of activity surrounding the development of transportation for LNG and oil exports to Asia, serves as a useful litmus test of the priorities of the government and the companies: hydrocarbon exports come first, regional development comes a distant second. Despite the lip service given by Putin and the government in Moscow to the development of the entire Arctic seacoast, in practice that development is largely limited to the oil and gas projects under way in the western part.

To put the future of Russian nuclear power in perspective, a global rebirth of nuclear power, if it happens, would not have a significant impact on the world's energy transition until the second half of the century. At mid-century—the focus of this book—the contribution of Russian nuclear power to solving the world's climate problems, and to offsetting Russia's own coming loss of hydrocarbons revenues, will still be modest. Playing the long game, as described in this chapter, will be the task of the next generation of Russian technologists and entrepreneurs.

In the meantime, Sergei Kirienko's star has risen ever higher. In 2016 he left Rosatom and was named first deputy head of the Presidential Administration. In 2018 he was awarded—by a secret decree—the order of Hero of Russia.[71] One might have expected him to be given the economic portfolio, but instead he was put in charge of domestic policy. Kirienko continues to win praise from Putin for his work in recruiting a new generation of political managers. We may not have seen the last of his surprising career.

Conclusions

Does nuclear power present an alternative model for Russia's future in the era of climate change? This question, as we have seen, has two aspects—domestic and international. The first is quickly answered. Overall domestic power demand is expected to grow by only 0.7 percent per year to 2050. The Russian domestic power market is, as one analyst puts it, "awash in capacity,"[72] and will likely remain so for the foreseeable future, so long as Russian economic growth remains slow, as it likely will. There will continue to be a small niche for new nuclear power plants, but after 2030 most of the domestic demand will consist of replacements.

The real question is the international market. As Roman Golovin suggests, the market is potentially vast. But the real question is whether Russia will be able to compete with China's increasingly aggressive export policy. In the near term, that question turns on whether Rosatom can keep innovating beyond its standby VVER 3+ series, and whether its fast-neutron model, BN-800, will be commercially viable. So far, the question remains open.

Whether there is a renaissance of nuclear power ahead as one of the world's answers to climate change depends on how one visualizes the future of energy overall. Advocates of wind and power imagine a post-fossil-fuel world of distributed power, in which autonomous hybrid systems, consisting of wind and solar plus storage, generate abundant clean electricity to supply local networks. Rosatom, with its recent initiatives in wind power, is making a side bet on this vision. Its foray into floating nuclear power plants is another side bet, which would also support distributed networks. Rosatom's interest in hydrogen generated by nuclear power is yet another side bet. But these remain a very small part of its total activity. Rosatom's DNA, one might say, remains the traditional large baseload nuclear power plant.

Large nuclear power plants will have their place in such systems, but the future of nuclear may be driven instead by small modular reactors (SMRs), which would have the advantages of greater flexibility, lower cost, and improved safety. Already several countries, including France, the UK, and China, have launched SMR programs. China has already built two small units that may begin production in the early 2020s. For the United States, SMRs could be the technology that would enable the Americans to get back into the global nuclear game. A possible milestone was passed in September 2020 when the US Nuclear Regulatory Commission issued its first-ever design approval for a small light-water reactor.[73] This new SMR, if it is actually licensed for construction, would be built from prefabricated modules transported by rail or truck—a radically different approach to nuclear power plant construction that might overcome the high costs and delays that have effectively killed new nuclear power in the United States. Russia, which continues to rely on classic large models, may find its current leading position weakening unless it can compete in the next generation of nuclear power.[74]

Yet the future of new nuclear technologies remains uncertain. This can be seen from the misadventures of another Rosatom program to develop small floating nuclear power plants.

7

RUSSIA'S AGRICULTURAL RENAISSANCE

The potential impact of climate change on the world's food supply is the most alarming of the dangers that may lie ahead. Certain regions, such as the Middle East and India, are especially vulnerable. Even the United States—the world's leading exporter of food—stands to be affected by climate change, with increased flooding in the Midwest, and droughts and declining water tables in the West. Yet some countries, including the United States, might also benefit, owing to increased demand for food exports and the enhanced geopolitical influence such exports could bring. How will Russia be affected?

At this moment Russian agriculture is booming. Over the past two decades, agriculture has become one of Russia's greatest success stories. For anyone who remembers the miserable performance of Soviet collectivized agriculture, and Russia's chronic dependence on imported food in the 1970s through the 1990s, this is an astonishing turnaround indeed. Russia is already the world's largest exporter of wheat, and its output is expanding rapidly across a wide range of foodstuffs. But will Russia's agricultural renaissance continue in the decades ahead, as climate change worsens? In this chapter we look at how this transformation came about, and what consequences it may have for Russian food security and income from exports in coming decades.

Stalin's Agricultural Disaster

Stalin plundered the Soviet countryside to drive the development of cities and industry. But the result was mass famine in the countryside, the destruction of the traditional village, and a legacy of impoverished and inefficient farms. Though food production improved somewhat under his successors, food distribution remained abysmal; only half of farm production reached consumers, while the rest spoiled in the fields. Collectivized agriculture was the albatross that hung over the Soviet system for five decades.[1]

Then came the Soviet collapse, bringing still more calamity. The impact on Russian agriculture over the next fifteen years was disastrous. Sown area dropped by one-third country-wide, from about 118 million hectares in 1990 to just under 75 million in 2007.[2] Even in the most productive regions of the south and southwest there were declines of 30 to 40 percent. In the so-called Non–Black Earth zone (in Russian, *Nechernozemʹe*)—a band of less fertile soils ranging from Pskov to Sverdlovsk, Tomsk, and Kemerovo—the amount of sown land dropped by half.[3]

How could such a history of misfortune have been reversed in the mere space of two decades? The answer is a massive change in ownership, consisting initially of wholesale privatization in the 1990s, followed by reconcentration after 2000 in the hands of more efficient producers, chiefly large private agroholdings.[4] Russian agricultural exports have flowed to world markets, while imports from Western countries have been restricted by the state in response to Western sanctions; in both cases the result has been a major stimulus to domestic production. Finally, both central and regional governments provide abundant subsidies to agriculture, although mainly to the largest producers.

This raises a series of fundamental questions about the future, as climate change worsens around the world. First, how solid is the Russian agricultural renaissance, and will it be helped or harmed as temperatures rise? Second, will Russian food exports continue to increase, and what revenues will they contribute? Thirdly, will food for the Russian population be secure? And lastly, what role will Russian agriculture play in the international diplomacy of climate change and the geopolitics of food?

The effects of climate change will be felt in two phases: during the first, running to the end of the 2020s, present trends will continue to drive increases in food production, exports, and revenue. But the following two decades, to mid-century, will bring a slow but steady erosion, although this might be offset by a rise in export prices.[5] This erosion will be due to a conjunction of six causes: externally, the evolution of global markets; internally, the steady worsening of direct effects from climate change; limits to the expansion of agricultural land; the longer-term effects of changes in land ownership; the impact of land reclamation policies; and finally, trends in financial aid from the state. We examine each of these in turn.

Russian Agriculture in Global Context

The story of Russian agriculture will play out against the backdrop of a deteriorating global environment. The worst impact of climate change is likely to be felt on the world's agriculture and food supply. The United Nations' Food and Agriculture Organization (FAO) warns that if global population reaches 9.5 billion by 2050—as it is currently on track to do—the world will need to increase food production by 60 percent compared to today. Yet the FAO projects that food supply could decline by 10 to 25 percent.[6]

Agriculture itself adds to climate change, accounting for about one-quarter of greenhouse gas emissions worldwide.[7] But it contributes to the coming crisis in other ways as well. Traditional agricultural practices in many parts of the world overuse water and abuse the soil. As a result, rivers and seas dry up, soils become laden with salt, ground erodes, and water tables sink deeper and can no longer support irrigation. In many places, fertility is already declining. These problems will worsen as climate change advances, an awareness that has led to

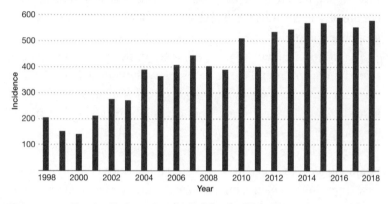

Annual Incidence of "Dangerous Weather Events" in Russia

Data source: Russian Hydrometeorological Service (Rosgidromet).

growing calls for wholesale reform of food production as well as deep changes in the ways the world processes and consumes food. But the inertia of the world agricultural system is so great that change will not come quickly.

What role does Russian agriculture play in climate change? Russia is the world's fourth-largest emitter of greenhouse gases (GHG). Agriculture alone accounts for 15 to 18 percent of Russia's GHG emissions, and together with the food industry it produces 25 to 28 percent of Russia's total, chiefly in the form of carbon dioxide and nitrogen oxide but also substantial amounts of methane from livestock.[8] This basic picture is unlikely to change because sustainable agriculture takes up only a very small share of total farmland and is unlikely to grow significantly any time soon.[9]

Russian agriculture will not escape the effects of climate change. Temperatures are rising 2.5 times faster in Russia than in the rest of the world. Between 1998 and 2018, there was a 300 percent increase in what Russian statistics call dangerous weather events. These include floods and fire, but the greatest threat is drought, combined with increasingly violent rains.[10]

Russian Perceptions and Initial Responses

These trends are causing Russian scientists and government officials to focus on changing weather patterns and their possible impact on food production. Yet on the local level there is still widespread skepticism about climate change as an explanation, particularly among farmers and local officials, and there is not yet any meaningful action to address it. For example, the commercial director of one large agroholding expresses a typical attitude, as he shrugs, "In some places it's worse, in some places it's better, but on average all the changes are within some sort of statistical error."[11] The major agroholdings report that their average yields over the last decade have been no lower than normal.[12] As far as the large farms are concerned, it all balances out. The statistics bear them out: between 2000 and 2018, average yields of grain and pulses have increased by half; and the last decade has been especially favorable. There has not been a major crop failure since 2012.[13]

Most players in the food sector do not yet believe climate change is real. Understandably, they have focused in recent years on Russia's own "agricultural miracle." Indeed, in the past two decades times have never been so good. The revolution in Russian food production is still gathering strength. Total grain harvests have nearly doubled since the end of the 1990s, and they are still growing. According to Stephen Wegren, the West's leading authority on the Russian food sector, "Russia has emerged as a grain powerhouse."[14] This is a very recent development: it was only in 2016 that Russian grain production returned to Soviet levels. But now grain exports in the best years are above 50 million tons. Revenues from agricultural exports, led by grain, now exceed $25 billion per year (wheat alone is about $8 billion), and President Putin has set a target of $45 billion by 2024. Export income from agriculture runs well ahead of arms sales (which Putin estimated in 2018 at about $15 billion per year),[15] and ranks third after oil and gas.[16] The outlook for exports over the next decade, at least on the eve of the COVID-19 pandemic, appeared bright.[17]

A major turning point in state agricultural policy came in 2014, when Putin responded to the West's economic sanctions after the annexation of Crimea by instituting countersanctions to ban many categories of food imports, notably from Europe.[18] The result has been a boom in domestic production, notably of meat and dairy products, which has reversed a long-standing decline in Russian livestock. Russians now pay higher prices for many types of food, but production is increasingly of domestic origin, and, as noted, food imports have declined steadily. For the foreseeable future, food security seems assured.

In short, there is not yet a sense of coming trouble, at least within the agricultural sector. But what will the longer-range impact of climate change be, particularly after the 2020s, and how will Russian farmers, scientists, and officials react to it?

Climate Change: Positive or Negative?

Climate change will alter the geography of Russian agriculture. A convenient way to think about the impact of climate change is to visualize a horizontal band, optimal for cultivation, that gradually moves north as temperatures increase. The southern boundary of the band is defined by increasing heat and drought, with negative effects on crop yields, especially in European Russia, where two-thirds of the territory is short of water. Grains, and wheat in particular, are highly susceptible to drought. The northern boundary is defined by increasing warmth and rain, with generally positive consequences. The net effect depends on how these three factors—warmth, rainfall, and length of growing season—combine. Russian models suggest that an increase of 1 degree Centigrade is likely to be positive for overall yields, but 2.5 degrees and above will be negative, and 3.5 degrees strongly negative.[19] In Siberia, where climate is more continental, the negative effects of climate change are likely to predominate.

Predicting the impact of climate change on Russian agriculture is complicated by the fact that there are so many different regional environments across the enormous expanse of Russia, and they will be affected differently by rising temperatures. Nevertheless, some broad generalizations are possible. In the north and east, the main problem will be excessive rainfall; in the south and west—the areas of greatest production of Russia's grain exports—the main problem will be drought.[20] Both will become more intense as the twenty-first century advances, particularly after the 2030s, with more frequent extremes, especially heat waves, throughout the country. More intense rains will lead to greater flooding, even in southern areas.

The impact on productivity will be especially severe in the south, where most of the country's grain exports come from. According to a 2018 report signed by Aleksandr Bedritsky, who was at that time the chief climate advisor,[21] yields (*urozhainost'*) in the grain-producing regions may decline by 5 to 10 percent by mid-century.[22] Some predictions are more dramatic. For example, academician Andrei Paptsov, a prominent authority on agricultural economics, draws from an extensive study by the Institute for Agricultural Meteorology to warn that grain yields will decline by 7 percent by mid-century and feed grains by 17 percent, while in the more marginal Volga and Ural regions the declines in the yields of feed grains will be 30 percent and 38 percent, respectively.[23] In contrast, the impact on more northerly regions of European Russia, especially the Non–Black Earth zone might be positive. These are areas that grow vegetables and fruits, and raise animals for meat, dairy products, and eggs, especially in clusters around the big cities. If their yields expand, the range and security of domestic supply could improve. One Russian journalist has joked, "We could have pears from Riazan' and apricots from Tula."[24]

From this basic picture, one might be inclined to conclude that Russian agriculture should be able to adapt successfully to climate change by moving its center of gravity to the north, opening up unused land in the north while expanding reclamation in the south. Adaptation becomes essentially a question of time and investment. But as we shall see in the next section, this notion of a moving band soon runs into a central problem: as one moves north, Russian soils become increasingly marginal.

The Central Facts of Cold and Poor Soils

Despite its enormous land mass—over 17 million square kilometers, or 1,700 million hectares—Russia paradoxically suffers from a shortage of land suitable for agriculture. The central reason, not surprisingly, is cold. Two-thirds of Russian territory is founded on permafrost, leaving only one-third as theoretically available for agriculture. But even within this category, cold remains a dominant factor. A recent authoritative study of Russian soils classifies 89 percent of agricultural land as cold—defined as zones with a year-round average temperature below 0 degrees Celsius. On this account, only 11 percent of agricultural land is warm enough for productive agriculture, and another 2 percent is only marginally available. Even within this zone, not all land is suitable; some estimates put the share of actual productive agricultural land at about 5 percent of the total land area.[25]

But as temperatures warm, won't the band of potentially fertile land expand to the north and east, compensating for whatever may be lost to heat and drought to the south? Russian climate and soil scientists warn that this picture is incorrect. As one moves out of the present agricultural land area, soils are generally poorer, thinner, and more acidic. To the north and east, even though permafrost will melt, the resulting exposed soil is infertile. Permafrost is not actually soil, but a mixture of sand and ice, and it has not had the long accumulation of humus that results from plant life or the action of underground organisms. A recent government survey of the state of Russian soils concludes flatly, "Russia has very limited resources of soil suitable for agriculture. Climate change will not increase this area. In other words, there is no potential for further expansion of agricultural land in the country."[26]

This phrase is key. Virtually the entire history of Russian agriculture has been based on "extensive" agriculture—the expansion of agricultural land by bringing new areas under cultivation. But what the scientists are saying is that that era is now over; there is no more prime land to expand to. In short, the outlook for Russian agriculture under climate change will depend on how successfully the country is able to manage its existing stock of land.

But everything depends on what one means by the existing stock. According to officials in the Ministry of Agriculture, there are more than 45 million hectares of unused and abandoned land that can be brought into production. Under present government programs, more unused cropland is being recovered every year (1.07 million hectares in 2019 alone), and the plan is to restore another 4 million hectares to production by 2025, mainly in the Non–Black Earth zone. As far as these officials are concerned, these lands are part of the existing stock. But climate and soil scientists disagree.[27]

One needs to put this debate in context. As a by-product of the Soviet collapse and the upheavals of the 1990s, a vast process of selection has taken place, as millions of hectares of land formerly under cultivation were abandoned or allowed to lie fallow. These were, by definition, the more marginal lands in each area. For example, large areas of central Russia, where grain used to be grown in Soviet times, were taken out of production or converted to other crops. The 2019 soil survey comments, "Over large areas, where the climatic and demographic conditions were unfavorable for the development of agriculture, enterprises in Soviet times were forced to plow up more land area than they were actually able to work and raise crops from."[28] The abandonment of this policy at the end of the Soviet period has been a favorable development for the productivity of Russian agriculture overall, as each region now specializes in what it does best. In the more productive southern regions of the Black Earth zone, grain output has increased strongly, in part because it is being conducted efficiently in the five Russian oblasts (provinces) best suited to it.[29] Grain production has become concentrated in these regions: 79 percent of the 2016 grain harvest was produced in south Russia and in the southern Volga basin, with only 11 percent coming from the Non–Black Earth region and 10 percent from southern Siberia.[30]

The danger of bringing unused and abandoned land into production wholesale, as state officials are seeking to do, is that it threatens to reverse this process of specialization, by investing money and effort into marginal areas. It amounts to a resumption of the traditional extensive approach to land management. In contrast, climate and soil scientists—as well as agricultural economists—advocate concentrating investment in precisely those places and crops where it will yield the most efficient and secure results. In effect this means breaking with the traditional industrial model of Soviet and Russian agriculture.

Confusingly, both sides call their recommended policy intensification (*intensifikatsiia*). But they mean two different things by it. For the Ministry of Agriculture, intensification means essentially adding more inputs—more land, more machinery, and more fertilizer and pesticides. For the scientists and economists, it means making the most judicious choice of crops and inputs on the basis of continued regional specialization. At the center of the debate is the role of reclamation and irrigation. These can be instruments either of an industrial model, if applied indiscriminately—as they were in Soviet times—or of a more sustainable model, applied where they will have the greatest impact on yields and stability.

The most critical test of these two models in coming decades will be the five southern oblasts that are the biggest and most efficient producers of grain, as well as the band of oblasts immediately to the north and east that are the next most productive zone. As droughts become more frequent and more destructive, how will the center of gravity of Russian grain production move, and which definition of *intensifikatsiia* will predominate?

The answers will depend on state policies, but above all on the responses of the farmers themselves. We turn now to the wholesale transformation that has taken place in the pattern of ownership of the farms and in their orientation to the market.

The Making of a Food Revolution, 1990–2020

The Russian agricultural renaissance consists of two parts—a sharp increase in exports and an equally sharp decrease of imports. One does not necessarily imply the other. Russia has become a major exporter of food; but it also remains a major importer. Over the last decade the balance between the two has swung heavily toward export. As recently as 2013, imports outweighed exports by $27 billion; but by 2020, at the onset of the COVID-19 pandemic, the balance was nearly even.[31] They do not involve the same products: the growth in exports consists mainly of grain and oilseeds; the decline in imports concerns mainly sugar, dairy products, meat, and processed foods. State policy at most times supports exports and discourages imports, as we shall see below. But the most important driver of change has been the response of the farms themselves, which have expanded exports in the south and replaced imports with domestic foods throughout the country. To understand what lies behind this response, we turn to the revolutionary change in ownership that has taken place in the last thirty years.

The structure of Russian agriculture today is a direct outgrowth of the radical reforms that were undertaken in the first years following the Soviet collapse, followed by an even more radical reconcentration after the mid-2000s, a powerful countertrend that continues to the present day. This took place in three phases. During the first phase, which began in the early 1990s, the most important single change was the transformation of the previous collective farms and state farms into jointly held "agricultural enterprises," in which farmworkers held land shares, which were leased to the enterprises. But as Stephen Wegren observed when commenting on the situation in the mid-2000s: "[This] brought the distribution of land shares, but not real land, to the vast majority of rural households. . . . Because land shares represent abstract and increasingly insecure property rights, in effect a very large portion of agricultural land is without a true owner. . . . The economic power of private property was never realized."[32]

Thus, through the mid-2000s, land reform had failed to remove most land from the control of large farms that were the descendants of the Soviet-era collective farms (*kolkhozy*) and state farms (*sovkhozy*). As late as 2007, 91.5 percent of the large farms were held in the form of land shares. At the same time, more than two-thirds of the large farms continued to be owned by the state and by municipal governments, which did not have the skills or the means to manage them effectively. As a result, the large farms went into a steep decline. By the end of the 1990s their output had plummeted to about one-third of the 1990 levels. Their productivity dropped while debt soared; arable land under cultivation diminished as large tracts of land were abandoned, and land reclamation stopped.

This was clearly a highly unstable situation. After 2000, when oil-export revenues started to put the Russian economy back afloat and the agricultural sector became profitable again, agricultural land started to attract the interest of outsiders, sometimes called raiders, who began acquiring the smallholders' shares on the cheap and taking over the large farms on which they worked. Large numbers of farmworkers were dispossessed in this way and were turned back into ordinary laborers. As a recent Russian study puts it:

> The privatized successors of collective and state farms were economically and financially weak, ripe targets for takeover; new legislation allowed market transactions in land and yet no barriers to excessive land concentration existed, which facilitated massive land accumulation; the rural population was poor and lacked entrepreneurial drive or business skills, often preferring to sell their land shares for quick cash rather than invest in farming.[33]

Or, as former agricultural minister Aleksei Gordeev put it more picturesquely:

> People with money arrive, with lawyers, and together with all the shareowners they sit in a bus, conducting registration. The shares are entered into some kind of auction and are bought. The new "owners" summon the director of the farm and say— goodbye, this land is no longer yours, and we have other plans for it.[34]

In parallel, a small minority of private farmers (*fermery*), who had taken the risk of separating themselves from the collective farms at the time of privatization, also became more stratified, as the more efficient (or better-connected) *fermery* expanded their holdings by buying out smaller ones. The result, within a few short years, was a bifurcated rural society in which a small number of winners coexisted with a large number of losers. For the latter, land reform was largely illusory. The main winners were the large farms, increasingly in the hands of outsiders and the more successful *fermery*.

The Rise of the Agroholdings

But that was only the beginning. Starting in the early 2000s, a third phase of the transformation in land ownership began, marked by the emergence of a new form of ownership, the agroholding. The timing was perfect: the devaluation of the ruble in 1998 made domestic farms an attractive target for anyone who had access to dollar revenues (for example, exporters of oil or metals), while rising international prices for agricultural produce made exports potentially profitable. The result was the agroholdings, an entirely new development in Russian agriculture.

Over the past twenty years, agroholdings have grown rapidly, absorbing smaller farms and diversifying into many different activities, including food distribution and retail outlets in Russian cities. Since the mid-2000s they have become favorite investments for banks, oil and gas companies, and other nonagricultural companies. The largest agroholdings are very profitable. During the ten-year period 2006–2016 their profits increased by three times, far outperforming the agricultural sector as a whole. By 2016 they accounted for more than half of the revenue from all "agricultural enterprises" (i.e., excluding *fermery* and household plots).[35] Not surprisingly, they are tied to the political / oligarchic elites of the country at both the regional and national levels. They are rapidly becoming the most powerful force in Russian agriculture.

The spread of agroholdings has been so rapid that until recently there was little comprehensive information about them. But a recent survey shows how quickly they have become major players. There are over 1,000 agroholdings, controlling more than 20 million hectares, or over 25 percent of all agricultural land.[36] The result is an increasingly concentrated production system: 20 agroholdings produce 49 percent of animal feed, while another 25 companies produce 43 percent of the nation's meat."[37] Only in the production of grain is the role of the agroholdings still relatively modest, although their share is starting to grow.

Agroholdings are such a recent phenomenon that they are not yet recognized as a legal form of ownership. Only since the mid-2010s, thanks to extensive research by the respected Russian Academy of National Economy and Public Administration (RANEPA), led by Vasily Uzun and a team of colleagues, has a full picture started to emerge. Their study found that by 2016 the largest 100 agroholdings, although they still controlled only 12 percent of all agricultural land, generated over one-third of all agricultural revenue. Since then the process of concentration into large agroholdings has turned into a tidal wave.

Agroholdings can be private or state-owned, but a striking trend in recent years has been the sharp decline in state-owned agroholdings—in number, size, and revenue. Between 2006 and 2016, the number of state-owned holdings dropped from 461 to 85, while the number of privately owned agroholdings more than tripled, from 315 to 978. In other words, state-owned holdings still predominated in 2006, but ten years later they accounted for barely 8 percent of the total. The numbers on sales revenue and profitability are even more striking. In 2006 the revenues of state-owned and privately owned agroholdings were about equal; by 2016 the revenues of privately owned agroholdings were 18 times larger than a decade before, whereas those of the state-owned holdings had hardly budged.[38] In short, the rise of the agroholdings is the result of a massive second round of privatization, at the expense of the rapidly shrinking state-owned sector.

THE NEW ELITE OF RUSSIAN AGRICULTURE

Who are the owners of the agroholdings? Agroholdings come in all shapes and sizes, and each one has its own unique history. Most of the early founders had modest beginnings, many of them outside agriculture. Some of them originated as pure traders, often importing nonagricultural goods. Thus Igor Khudokormov, the founder of Prodimex, the largest agroholding today in terms of total land under its control, began in the early 1990s by importing sugar from Ukraine. Likewise, Vadim Moshkovich, the founder of Rusagro, the country's third-largest agroholding, also began as a trader by importing sugar, although he had previously dabbled in oil and vodka. Today Prodimex and Rusagro both control 700,000 to 800,000 hectares of land and are Russia's largest producers of sugar beets and sugar. Another example is Miratorg, founded and owned by a pair of twins, Viktor and Aleksandr Linnik, who started out importing dried milk from Holland.[1] Toward the end of the decade they began importing meat from Brazil, and in the mid-2000s they made the switch to raising animals. Today Miratorg is Russia's largest producer of beef and poultry. None of these founders has a background in agricultural science. For those founders who had some previous connection to food, the connections were mostly indirect. For instance, Igor Babaev, the founder of Cherkizovo, a leader in pork and poultry, had spent his entire earlier life in the meat-processing industry. By 1989 he was director of the Cherkizovo meat-processing factory, which he moved quickly to privatize as soon as Gorbachev's reforms allowed. Cherkizovo has grown from that early base.

In some instances, well-known personalities from other industries have started second careers as owners of agroholdings. Sergei Kukura was one of the founders of LUKOIL Russia's first private oil company.

1. The story of the Linnik twins is even more picturesque. They started out as engineers in the defense industry, but quit because it was poorly paid. They got involved in food through contacts made while providing travel services for tourists. One suggested importing dried milk from Holland, and offered to stake them. Before long the brothers were importing large quantities of food. See "Interview with Viktor Linnik" in *Agroinvestor*. Reprinted in *RusLetter*, October 24, 2014, http://rusletter.com /articles/the_onset_of_the_pigs_how_miratorg_conquered_the_russian_food_market.

After twenty-four years with the company, during which he became deputy general director and CFO, Kukura sold his stake in LUKOIL and created an agroholding, Volgo-Don Invest. Based largely in Voronezh, it is already the fifth-largest farmland owner in Russia.[2] Kukura is the classic case of the outsider—a lifelong oilman, he had no prior experience in agriculture. Now in his late sixties, he has turned over the management of Volgo-Don to his son Alexei, one of a growing number of examples of second-generation owners of agroholdings.

One of the few true insiders is Aleksandr Tkachev, whose father was the chairman of a collective farm in Krasnodar, which he privatized early under the name Agrokompleks. Agrokompleks is still based in Krasnodar, where two-thirds of its 670,000 hectares are located. Agrokompleks produces a wide assortment of foods, which it sells through a network of more than 700 retail stores throughout Russia.[3] Tkachev was governor of Krasnodar for fourteen years (initially as a member of the Communist Party and subsequently as a member of the progovernment United Russia Party), and then minister of agriculture from 2015 to 2018. Agrokompleks prospered in parallel with Tkachev's rise in politics, leading to periodic accusations of conflict of interest. In 2015, when Putin named him minister, Tkachev divested himself of his stake in Agrokompleks in favor of his relatives, who had long been co-owners of the family business.[4] By the time of Tkachev's appointment as minister, the surge of agroholdings and exports was already well under way throughout the country. But Tkachev left a lasting mark on policy by pushing food security and protectionism, and was a strong supporter of Russia's food "counterembargo," which stimulated production for the domestic market.[5] Not surprisingly, he is known as one of Putin's strongest supporters.

2. LaScaLA, *Top 10 Russia's Largest Agricultural Landholders 2018,* https://www.largescaleagri culture.com/home/news-details/top-10-russias-largest-agricultural-landholders-2018/.

3. Interview with Evgeny Khvorostina, the general director of Agrokompleks, in *Agroinvestor,* July 17, 2020, https://agrovesti.net/news/corp/agrokompleks-im-tkacheva-investiruet-v-rasshirenie -proizvodstva.html.

4. "Pravitel'stvo zavershilo proverku biznesa sem'i Tkacheva," Slon.RU, April 19, 2016.

5. By the time of Tkachev's departure in 2018, Russia was reported to be close to self-sufficiency in a wide range of products—93 percent for sugar, 84 percent for vegetable oils, 90 percent for meat, 97 percent for potatoes, and 82 percent for dairy products. "Aleksandra Tkachev otpraviat v otstavku s povysheniem," Sfera.FM, May 7, 2018, https://agrovesti.net/news/indst/aleksandra -tkacheva-otpravyat-v-otstavku-s-povysheniem.html.

There are few foreigners among the founders of the top agroholdings. Technically, foreigners are barred by law from owning land in Russia, but according to a recent survey foreigners own as much as 5 percent of all agricultural land and generate over 16 percent of all revenue.[6] Many of them are based in Cyprus, which suggests that the actual owners may be Russians, but because there are no official statistics on foreign ownership, it is not possible to be certain. One of the few clearly identifiable foreigners is Stefan Durr, the German-born founder of EkoNiva, Russia's largest producer of milk and dairy products. Durr first came to Russia in 1989 as a student and worked on a pig farm, then went into the milk business in 2005. In 2014 President Putin recognized his achievements by granting him Russian citizenship, but Durr is also an important dairy producer in Germany, and he is held up as a symbol of successful Russian-German relations.[7]

6. Interview with Vasily Uzun in Tat'iana Bogdanova, "Ch'ia v Rossii zemlia," *Argumenty i fakty*, February 5, 2020, accessed via East View (https://dlib.eastview.com/).

7. For a profile of Stefan Durr see Howard Amos, "The True Tale of Russia's Dairy King," *Moscow Times*, November 23, 2015.

The new elite of Russian agriculture represent a mixture of origins and personalities (see the box above). There are as yet no dominant personalities who tower over the whole field, such as Oleg Deripaska in aluminum or Leonid Mikhelson in LNG, or the Kremlin nominees who preside over the large quasi-monopolies, as in oil and gas. At present the growth of the agroholdings is like a turbulent pot at a high boil, reminiscent of the 1990s in the energy and metals sectors, although not nearly as violent. The rankings change every year, and many players do not survive. Of the 100 largest agroholdings identified in 2006, 46 had been liquidated ten years later and another 13 had exited from primary agricultural production. Many ambitious plans for expansion, initially announced with great fanfare, have never been realized.[39] But these are still early days, and further concentration appears inevitable. What will it mean for Russian food security and food exports, as climate change intensifies?

Until recently, agroholdings played only a modest role in grain production and exports. According to the 2019 RANEPA survey, in 2016 the share of agroholdings in total cereals production by all farms was only a little over 20 percent. The main producers of cereals and oilseeds were still independent enterprises and family farms (a category consisting mainly of *fermery* and some private plots). For the time being, *fermer*-owned farms have continued to prosper; for example, in 2017 they produced 30 percent of all cash crops, such as grain, sunflower, and flax, and are thus still major providers of exportable output.[40]

But agroholdings are now starting to enter the grain business, as the growing profits from grain exports draw the interest of investors from other corners of the economy. The leading example is the company Steppe (in Russian, *Step'*). Steppe is a diversified agroholding that produces milk and other dairy products, but it is increasingly active in grain production as well as exports, investing in grain elevators and port facilities, thus combining the entire value chain. Although so far it has captured only about 3 percent of the grain export market, Steppe, significantly, is part of the powerful Moscow-based conglomerate Sistema, which is majority-owned by one of Putin's closest allies, Vladimir Evtushenkov.

Some of the traders who buy from independent *fermery* are well-known international brands, such as Cargill and Glencore, but the share of foreign traders has declined sharply in recent years, in favor of rising Russian players.[41] The top Russian grain trader, with 25 percent of the market, is the company RIF, based in Rostov, the creation of a local self-made businessman, Petr Khodykin, who built the business from scratch. Khodykin is a pure trader, and he plays no part in production. Khodykin's strategy is to focus on logistics. He has bought thousands of hopper cars and hundreds of river-to-sea ships, which enable him to reach deep into the countryside and buy from farms that others cannot get to. So far this approach has been successful, and RIF's share of the grain export market has been growing rapidly. However, Steppe is a serious competitor to RIF's *fermer*-based business model. And whereas Khodykov presumably enjoys the support of the Rostov Oblast government, it may be hard put to compete with the muscle of the Kremlin. Steppe, not RIF, may represent the future of grain exports.[42]

The advent of Moscow-based conglomerates as major grain producers and exporters could have major implications for the longer-term resilience of Russian grain exports under climate change. Unlike the local *fermery* which have dominated the grain trade until now, a conglomerate such as Sistema has much greater access to financing for investment in land improvement and the restoration of abandoned land. Yet it will also be under more direct pressure from the Kremlin to expand production and exports. Russian soil scientists have already raised concerns that the overenthusiastic development of grain production may ultimately harm the land.[43]

The implications of private agroholdings for the future of Russian agriculture are as yet unknown, but much depends on the choices they make in the coming decade. For the present, they are focused on expansion—horizontal expansion of the land under their control, and vertical expansion to cover the entire value chain from the field to retail stores, or, in the case of grain exporters, from the field to the export destination. Their investment is going primarily toward the activities that support these expansions, such as modernizing storage and processing—critical activities in a sector in which traditionally, in Soviet times, only half of the output from the field ever made it to the store—or expanding port capacities for exports. In the fields, the agroholdings are sharply increasing the use of machinery and the application of fertilizers.

The key question ahead is how the new owners will respond to increased damage from climate change—chiefly in the form of drought and flooding—and whether they begin to accept the idea that long-term adaptation is necessary to protect their lands. Overall, during the last decade they have neglected the problems of soil degradation and have as yet done little to protect the long-term fertility of their new holdings. The chief source of their increased productivity has been the application of fertilizers. Soil scientists worry that fertilizers are being overused. The result is widespread acidification, which according to the Ministry of Agriculture has increased sharply over the last twenty years and now affects 30 percent of all Russian cropland.[44] There are other problems, such as soil erosion, which at present are also not being addressed.[45]

In short, will the new owners take care of their land? Adapting to climate change will require increased investment, particularly in irrigation, drainage, and flood control. One question for the future is the rate at which the agroholdings will invest in information technologies, for purposes such as optimizing the application of water and fertilizers in the fields, reducing waste throughout the food chain, monitoring export prices, and controlling inventory in their retail outlets.[46] The leading agroholdings are just beginning to monitor the state of their crops and herds by using computers, GPS systems, and even drones.[47] The systematic application of information technologies could be a major factor in improving the ability of the agricultural system to respond to climate change.

At the present stage of their growth, the agroholdings as a group are not financially strong enough to undertake major projects in reclamation and soil improvement. They have been spending the bulk of their resources on land acquisitions, most of which are acquired on credit. According to Georgy Safonov, a noted agricultural economist, the agricultural sector is currently over 4 trillion rubles (approx. $57 billion) in debt,[48] with much of it owed by the agroholdings. In this situation, support for climate adaptation will have to come from the state. How much support is available and how is it allocated, and for what, and to whom?

The Role of the State in Agriculture under Putin

The reform of agriculture over the last thirty years has been a story of constantly shifting policies, in which the successive phases were driven by the influence of whatever group of policymakers happened to be dominant at any given moment. In the early 1990s the radical market reformers attempted to promote a new Russian agriculture based on smallholders. As we have seen, this policy largely failed. Then in the later 1990s regional authorities gained the upper hand and allowed the initial privatizations to be weakened or reversed, while in Moscow policy was paralyzed by political battles between conservatives and liberals competing for the ear of a weakening President Yeltsin.

The recent renaissance of Russian agriculture coincides with a third phase of state policies, adopted under Putin. Putin put food and agriculture near the top of his agenda from the moment he took office.[49] His first step was to stabilize the agricultural base, especially the larger farms on which most of agricultural output depended. Putin sharply increased the flow of state credits and loans to the sector, notably by creating a dedicated state-owned agricultural bank. He simplified the tax system and lowered taxes on the larger farms. He initiated a program of debt relief, which restructured the debts of the larger farms. He introduced a crop insurance program, and instituted price supports for key crops. State agencies purchased farm machinery and leased it to farms at subsidized rates. The government began intervening in grain markets to stabilize prices. The result was a sharp improvement in the farms' financial condition: whereas in 1998 nearly all the larger farms had been unprofitable and were going into debt, by 2008 the number of debtors had dropped to 25 percent, although since then farm debts appear to have risen again, partly as the result of the agroholdings' credit-fueled land acquisitions.[50]

The state's policy of supporting agriculture was strengthened further during the second decade of Putin's rule. Over the decade between 2006 and 2015, federal funding for agriculture increased eightfold, with steady increases every year, although the onset of recession in 2015–2016, followed by COVID-19, interrupted the upward trend.[51] Yet the state continues to subsidize a wide range of inputs, including fertilizer, credit, farm equipment and machinery, animals, seeds, and services,[52] and actively supports the development of new crops, such as soybeans for animal feed.[53] The state's support for Russian producers of agricultural machinery, combined with mandatory rules on import substitution, has enabled domestic manufacturers to capture nearly 60 percent of the market, compared to 25 percent in 2012.[54]

Yet the agricultural sector remains deeply affected by the long-term aftereffects of the Soviet collapse. Mechanization is a case in point. Farm machinery is antiquated; the average tractor is more than ten years old, and machinery is depreciating faster than it is being replaced. The domestic farm machinery industry has not yet fully recovered from the impact of the disastrous 1990s, when the annual production of grain combines, for example, fell from 66,000 to 1,000, and even today domestic factories supply only about one-third of the sector's equipment needs.[55] In short, the revival of the agricultural sector is still a work in progress and will require strong investment support from the state for the foreseeable future.

But so far the state's commitment to agriculture remains strong, as it has for the past twenty years. The revival of the food sector has been one of the signal achievements of the Putin era, a point that Western coverage of Russia has not adequately recognized. However, three key points should be noted here. First, when measured in terms of constant rubles, the increase in state support for agriculture has been far more modest than appears at first. Despite the extraordinary increase in state support in nominal rubles ("rubles of the day"), in constant 1995 rubles actual budget expenditure on agricultural support in 2017 was only about half the level of 1995.[56] Inflation has eaten away most of the real increase.

The second point is that Putin's policy has consistently favored the larger farms, and this is what has catalyzed the rise of the agroholdings since the mid-2000s. This imbalance has caused considerable unease and criticism among Russian experts.[57] As Uzun and his colleagues at RANEPA observe bitingly:

> Approximately 1% of all agricultural enterprises are selected by some opaque mechanism and receive the bulk of state support, with large agroholdings always at the top of this list. Government bureaucrats, guided by a traditional ideology of economies of scale, which has no foundation in empirical facts, freely distribute huge sums of state funds to a limited number of private entrepreneurs. . . . The strong large-firm bias in the distribution of state support explains why agricultural growth in Russia is indeed driven by the largest agroholdings.[58]

The third point is that most of the state's support, and much of the effort by the agroholdings themselves, has been devoted to near-term measures—financial consolidation, mechanization, enlargement of roads and ports, and so forth. Much of it, indeed, consists of short-term finance to help the farms get in the crops. Little attention is being given to the long-term condition of the soils themselves, and to the coming challenges of adapting to changing weather conditions, particularly drought and flooding.

The dominance of the short term over the long term in the state's policy can also be seen in the chronic weakness of agricultural science and extension services to the farms, a situation that has changed little since Soviet times. In a recent high-level meeting of the Security Council, former president Dmitry Medvedev (now in the more modest position of deputy head of the Council) called the situation "appalling." His critique echoed many of the characteristic weaknesses of the Russian R&D system: low priority and lack of funding, but above all a lack of communication between the research institutes and the ultimate users. Medvedev concluded: "Money is needed for scale-up, for advertising, and for extension services. But the scientific organizations get money only to develop new varieties—and very little at that—and while there may be good hybrids among them that could compete with the West, unfortunately no one knows about them."[59] As a result, most seeds are imported from the West, and little attention is given to developing new varieties that will be adapted to local conditions.

Reclamation Is the Key

Over the longer term, the sustainability of Russian food production will depend on whether the state now takes a longer-range view, chiefly by increasing spending on reclamation, and whether the agroholdings themselves prove to be caring, long-term custodians of the lands they now control.

They will have their work cut out for them. In the last two decades of the Soviet era the government deployed a large-scale program of reclamation, but much of the effort and money went to waste;[60] and in the 1990s the policy was abandoned as the country went through near-chaos. Only in the 2000s did Russia turn once more to land improvement. But the inherited system is in bad condition. As a result, irrigation and drainage systems are badly outdated or nonexistent. Where irrigation systems do exist, they operate only on 3 to 3.5 million hectares of the total 8 million hectares of cropland under cultivation. The area served by drainage systems is about the same.[61] The systems are aging, and only about half of the irrigated area is considered to be in "satisfactory" condition, while the state of the drainage systems is even poorer, with only about 14 percent considered satisfactory.[62]

Most of the reclamation system remains state-owned. The Ministry of Agriculture's Reclamation Department controls an empire of more than 40,000 reservoirs, dams, and canals.[63] In recent years the state has provided an average of 20 billion rubles (currently about $270 million) per year for reclamation, of which about 80 percent comes from the federal government and the rest from regional governments. Of that amount, about one-quarter goes to capital investment, the rest being subsidies, chiefly in the form of credits.[64]

These are modest sums, compared to the vast amounts the agroholdings are spending themselves, chiefly to acquire new lands.[65] Since

2005 there has been no net increase in the areas under irrigation and drainage.[66] Former president Medvedev singled out this problem, too, when he called the condition of the reclamation system in many regions "a deplorable spectacle, degraded and looted." Over 70 percent of the reclamation system requires major modernization, Medvedev concluded, and he called for sharp increases in state investment,[67] but this has not yet happened.

The Long-Term Impact of Climate Change

In sum, how will climate change affect Russian food production by mid-century? There are three outcomes to consider: overall production, food export, and domestic food supply. Each of these is likely to be affected in different ways.

First, one must pay tribute to the renaissance that has taken place in Russian agriculture over the past twenty years. One reasonably safe prediction is that the present momentum of change in ownership and investment will enable Russia to continue increasing production, both for exports and for the domestic market, for at least another decade.[68]

The question is what will happen after 2030, when we can anticipate more frequent and severe drought, heat, and violent rain. This would logically have the effect of moving the optimum band for agriculture to the north. Yet there will be little room for expansion northward, in view of the poor quality of soils, even where permafrost has melted. In other words, one can visualize an increasingly narrow agricultural zone, squeezed by heat and drought to the south and constrained by floods, excessive rainfall, and poor soils to the north, only partially offset by greater warmth and longer growing seasons. As a result, a growing share of Russian soils will become marginal.

But that is only half of the question. The other half is the capacity of the system to adapt. Here the two major variables are the effects of the change in ownership that has taken place over the last twenty years, and the level and effectiveness of state support.

The major new development, as we have seen, is the rise of agro-holdings as the dominant form of landholding. These have caused a sharp rise in short-term productivity, mainly owing to the increased use of fertilizers and imported inputs such as seeds and machinery. The main impact of agroholdings so far has been in the non-grain sector, particularly poultry and pork, which have grown vigorously, but the agroholdings are now penetrating the grain sector as well, with similar results. The rapid spread of agroholdings is the chief reason Russian food production and exports will likely continue to grow for another decade.

However, there are serious potential weaknesses to this structure. The first is its present near-exclusive focus on the short term. The agro-holdings are almost entirely concentrating on expansion. As a result, they are deep in debt, as is the agricultural sector as a whole (although the grain-exporting regions appear to be an exception). Little money is being invested in reclamation or other forms of longer-range adaptation. Unless this changes, and the agroholdings adopt a longer-range perspective, they will be vulnerable to extreme weather as climate change worsens.

The second weakness is the pattern of state support. The central ministries and their regional offices give major subsidies to agriculture, but most of the state's financial support goes to the largest farms, chiefly the agroholdings. Very little of it is earmarked for reclamation. Even where reclamation is being supported, the agricultural bureaucracy in Moscow is focused on bringing abandoned lands and other marginal croplands back into production.

In short, the existing pattern of state support reinforces the traditional industrial model of extensive agriculture, in which the main focus is on boosting inputs such as mineral fertilizers and pesticides, rather than increasing the efficiency with which they are used. Sustainable agriculture gets lip service in Moscow, but it runs counter to the industrial model, which is so deeply embedded in the system that it is unlikely to change. When confronted with inadequate rainfall, or conversely massive floods, the system's response is likely to take the form of crash programs in irrigation and water management on an industrial scale, just like in the 1970s of the Soviet period, but with equally little result.

As climate change brings more frequent droughts and heavy rains, the burden on state programs will increase. The Putin era has brought food security and growing export revenues, but in coming decades important changes in policy will be required. In addition to measures to strengthen the resilience of the existing land base, another key measure will be expanded insurance coverage against losses from extreme weather events, and assistance in recovering from them, to protect the financial health of the agricultural sector. In both cases a longer-range outlook will be needed, beginning with a greater recognition by all sides of the reality of climate change.[69]

All this would lead to a pessimistic outlook for the future of Russian agriculture by mid-century, were it not for a few additional factors. The first is that other countries will be suffering too, especially the United States and Australia. The result may be a worldwide contraction of food supply, which will drive prices higher, preserving Russia's export revenues even if volumes shrink. The contraction will be especially pronounced in developing countries, particularly the Middle East, which may open new opportunities for Russian geopolitical influence via food aid.[70]

How will COVID-19 and climate change affect the balance of Russian agricultural trade? In the short term, the effects of the virus are likely to be considerable; in particular, it may well prevent the government from reaching its target of $45 billion in exports by 2024.[71] By the late 2020s, however, the effects of COVID will fade, and the impact of climate change will become dominant, especially in the southern regions that generate most exports.

In sum, as Russia approaches the middle of the century, the food system will be under growing stress. As overall revenues from energy exports decline, especially from fossil fuels, the state will have fewer resources to intervene as it has done under Putin. The Russian farms and their owners will increasingly face the global market, and their own problems, alone. At that point, the resilience of Russian agriculture will depend on whether the Putin decades have truly strengthened the countryside, or turn out to have been a brief rush of prosperity and profit-taking.

8

A TALE OF TWO ARCTICS

On May 29, 2020, a giant storage tank for diesel fuel on the site of a power plant in Norilsk, the capital of Russia's nickel industry, abruptly collapsed as its cement foundations gave way, spilling 21,000 tons of diesel into the nearby Ambarnaia River. The fuel turned the river bright red as it ran toward the Arctic Ocean, spreading pollution that would likely poison aquatic life for the next decade. The investigation that followed revealed that the storage tank was more than forty years old and, like many Soviet-era structures in Norilsk, had not been maintained,[1] and had been undermined by the melting of permafrost under its foundations. The spill was comparable in size to the 1989 *Exxon Valdez* accident in Alaska.

The incident touched a nerve in Russian public opinion, and for the next several weeks media coverage was intense, both in newspapers and on social media, as thousands of people vented their anger. President Putin reacted in fury, and conducted a series of highly publicized online videoconferences, demanding that those responsible be identified and punished. The plant's director and two top engineers were arrested and charged with negligence, as Moscow officials descended on Norilsk and Putin declared a federal state of emergency. In the direct line of fire was one of Russia's earliest and richest oligarchs, Vladimir Potanin, the main owner of Nornikel', the company to which the collapsed tank belonged. Putin publicly dressed down the unfortunate Potanin, who pledged to devote part of his personal fortune to the

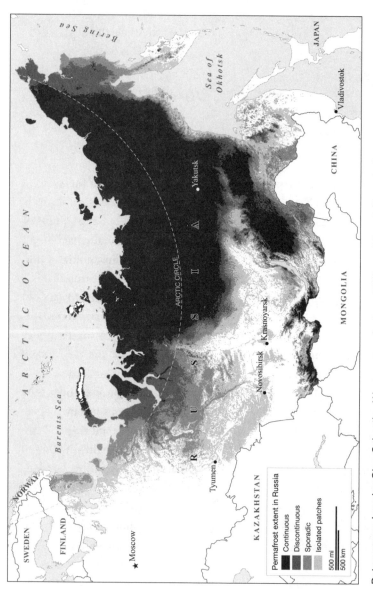

Data source: Jaroslav Obu, Sebastian Westermann, Andreas Kääb, and Annett Bartsch, Ground Temperature Map, 2000–2016, Northern Hemisphere Permafrost. Alfred Wegener Institute, Helmholtz Centre for Polar and Marine Research, Bremerhaven, PANGAEA, 2018. https://doi.org/10.1594/pangaea.888600.

cleanup, but even that did not exhaust the president's public anger, and he continued to pound away on local officials well into the summer.

For locals in Norilsk, the diesel spill was nothing new. In the wake of the accident a Russian correspondent interviewed a thirty-year veteran construction engineer at Norilsk, who had this to say:[2]

> Cries about damage to the environment are fifty years too late. . . . The rivers and lakes and the soils of the tundra at Norilsk have been [poisoned] for a long time now. Those wastes are a thousand times more harmful than diesel fuel. What kind of spawning grounds are they talking about? It's been decades since anyone caught fish there. The water smells like chemicals; if you eat fish from there, your next stop will be the morgue.

But the real significance of the Norilsk episode lay in what it presaged for Russia's future. The city is built on permafrost, a mixture of sand, soil, and ice[3]—the same mixture that covers two-thirds of Russia's landmass. As it melts, the entire infrastructure degrades. Many saw the spill at Norilsk as a wake-up call for the entire Arctic region of Russia. Months later there were still hundreds of articles in the Russian media about the consequences of the spill.

From this episode one might conclude that problems caused by melting permafrost, and the increasingly urgent need to invest in adapting to it, loom increasingly large in the minds of Russian policymakers, including the president himself. But that is not the case. In reality, there are two opposing narratives about the Far North. The first, largely supported by the scientific community and the local population, calls for large-scale investment in infrastructure throughout the Siberian inland to enable it to withstand the effects of global warming. The other narrative sees global warming as an opportunity. As sea ice melts along Russia's northern coast, it opens up the prospect of a major new seaway to Asia and stimulates the development of the entire Arctic coastline.[4]

Russian policy clearly favors the second narrative. The Kremlin's priority is to focus on the sea route and use it to develop and export oil and liquefied natural gas (LNG). In contrast, the vast landmass in the interior is being neglected. Given the constraints of cold, distance, poor soils, and declining population in much of the Arctic region, it is a logical choice, driven by Russia's comparative advantage in exports of raw materials, especially hydrocarbons. But over the long run the result will be to increase even further the specialization of the Far North and East around a handful of tradable resources, even as the long-term value of those resources may be decreasing. At the same time, the social and economic costs of neglecting the Arctic inland will be increasingly severe. So far, however, there is little sign that the Kremlin acknowledges this dilemma, or even perceives it to be one.

In this chapter I examine the two contrary trends currently under way in the Russian Arctic—the strong development of the Arctic coastline, led by oil and gas development, and the contrasting demographic and economic decline of much of the inland. The two big questions for the future of the Russian Arctic are these: How fast and how far will the natural-resource development of the Arctic coastline proceed as the coastal ice melts and the Northern Sea Route opens up, and what will the consequences of climate change be for the inland, from the accelerating melting of permafrost?

The Coming Decline of the Arctic Inland

In many ways the Arctic defines Russia. About one-fifth of Russian territory lies north of the Arctic Circle, but that fact alone does not tell the story. Some two-thirds of Russia rests on permafrost; as the climate warms, the permafrost is melting faster. Mean temperatures are rising 2.5 times as fast in Russia as in the rest of the world, but in the Russian Arctic they are rising five times as fast, and in some parts, seven times.[5] The harmful impact on human health and habitat is already severe.[6]

Much of Russia's natural-resource wealth is concentrated in the permafrost zone. In European Russia, oil, gas, and coal are produced in the Pechora basin of the Komi Republic. Beyond the Urals, in West

Siberia, some of the world's largest reserves of oil and gas were developed starting in the 1960s and have now entered a new generation, based on the Yamal Peninsula. Farther east, in Krasnoyarsk Kray, huge deposits of nickel and other metals are mined at Norilsk. In Sakha, at the northeastern end of Russia, there are major centers of coal mining and metallurgy, as well as mining of diamonds. In contrast, the central regions of European Russia, where the population, agriculture, and manufacturing are concentrated, are poor in natural resources and rely on long-haul transportation of fuel and raw materials and long-distance power transmission from the north and east to the west.[7] The Arctic was the Soviet Union's treasure house.

What made the Soviet model of arctic development possible was the unlimited supply of manpower. Most of the industrial resources of the Arctic north and east up to the 1960s were developed by convicts, concentrated in permanent settlements, which then became mining towns. These remain in place today; north of the Arctic Circle there are forty-six cities and towns with over 5,000 residents (in contrast, in Canada there are none).[8] When seen from space at night, Russia has a characteristic signature: in contrast to Canada, the north and east are lit by bright points of light, indicating the location of major urban clusters.

Russia no longer has the manpower—or the repressive political system—to support this model of development. Many of the older mining centers are in decline. Only about 10 percent of the population, or 13 million people, live in the permafrost zone,[9] and the number is shrinking, as people leave the Arctic inland and move toward the cities farther south.[10]

Melting permafrost is accelerating this process. The Russian towns along the coast of the Arctic Ocean are even now suffering extensive damage. More than 60 percent of the buildings in Igarka, Dikson, and Khatanga are cracked and deformed by the melting ground, over 50 percent in Pevek and Anderma, and 40 percent in Vorkuta. In the remote villages of the Taimyr region in Krasnoyarsk Kray, the figure is close to 100 percent.[11] Railroad tracks twist apart, roads heave, and pipelines rupture. Most of the road system in the Far North consists of ice roads, which are usable only in winter; as the cold season shortens, the region becomes more isolated. An increasingly serious consequence of melting permafrost is coastal erosion. At present the Russian coastline is retreating by one to five meters per year, in some places by as much as ten.[12] In coming decades rising ocean levels will increase erosion all along the 5,500-kilometer coastline of the Arctic Ocean, further destabilizing buildings and industrial structures. In the East, where the consequences of climate change have been especially pronounced, melting ice, combined with more powerful rainstorms, has led to major flooding, as rivers overflow their banks and cover hundreds of square kilometers on either side.[13] The traditional way of life of Indigenous Peoples—the Khanty, the Mansi, and the Sakha, nomads whose livelihoods depend on reindeer herding, hunting, and fishing—is increasingly threatened.

Not all of the Arctic inland is declining. Yakutsk, with a population of around 340,000, is the second-largest city in the Russian Arctic, after Murmansk.[14] But unlike Murmansk, which enjoys the warming influence of the Gulf Stream, Yakutsk, the capital city of the northeastern republic of Sakha, lies in the coldest region of Russia. Even so, Yakutsk is booming. From 186,000 inhabitants in the last Soviet census in 1989, the city's population had nearly doubled to 338,000 by 2018.[15] The explanation is the flip side of the depopulation of the Arctic inland: a steady exodus is under way from the desolate countryside, as ethnic Sakha flock to the city. The good news is that this has made Yakutsk, in the words of two leading ethnographers, "into a genuine Indigenous regional capital, the only one of its kind in the Russian North."[16] As its population has swelled, Yakutsk has become an administrative, political, cultural, scientific, and economic center, unique in the region.

But even Yakutsk cannot escape from climate change. Even though it lies south of the Arctic Circle, it too is mostly built on permafrost. As global temperatures rise and the ice melts, the foundations of Yakutsk are becoming unstable. Buildings there are traditionally raised up on piles driven into the frozen ground, but the load-bearing capacity of the permafrost has decreased by as much as half since the 1960s, which makes buildings vulnerable to cracking and ultimately to collapse. Yakutsk is a preview of the future of the Arctic inland: a handful of large centers in the midst of an increasingly depopulated—and liquid—northern tundra.[17] The long-term threat to the region is that much of the traditional economic basis of the Arctic inland becomes unsustainable.[18]

Clearly much of the response to the problem of melting permafrost must consist of reinforcing existing structures and buildings, and improving roads. New industrial buildings must be built on deeper piles; ports will require seawalls; and pipelines must be raised on struts and cooled, to prevent oil and gas, which are hot as they come out of the ground, from melting the surrounding soil. Dikes must be raised against flooding. Power lines must be reinforced. All of this involves extra costs, and it is at best a rearguard battle.[19] So far the response of the authorities has been weak, and Moscow has been mostly indifferent.

What role is Russian public opinion playing? At present the domestic political reaction to the warming of the Russian Arctic, with the exception of passing episodes such as the spill at Norilsk, remains confined to the local populations. However, within this group environmental activism has become, understandably, a growing political cause. Moreover, thanks to the internet, the concerns of Arctic communities are being displayed for all Russians to see.[20] As climate change gains in intensity as a political issue throughout Russian society, the melting of permafrost and its impact on the demography of the North, especially on Indigenous Peoples, could become part of a larger political movement against the unbalanced development of the region. But so far public opinion has not been a significant force for change. As one recent study concludes, "The majority of the population considers climate and environmental changes locally, does not associate them with global drivers, and is not prepared to act on them. Accordingly, even the best designed climate policies cannot be implemented in Northern Russia, because there is no public demand for them."[21]

Meanwhile, Moscow's attention is focused on the benefits that will come from the opening up of the northern coastline, to which we turn now.

The Expanding Alternative: Development by Sea

The historic dilemma of Russian leaders is this: If one looks at a map of Russia, there are broadly two possible answers to how to develop the Arctic. One is to proceed by land. Beginning in the late nineteenth century, Russia shipped goods and people eastward by rail, first by the Trans-Siberian Railway (and more recently also by the Baikal-Amur Mainline), and then north by Siberia's powerful rivers, the Ob, the Lena, and the Yenisey. That is how most of Siberia was developed. The other answer was to move east by sea along Russia's Arctic coastline, developing the rich natural resources along the way through a series of coastal outlets. The second answer, however, was not possible beyond the Yenisey River, because solid ice blocked the eastward route.[22]

Now the dilemma is reversed. As climate change makes existing towns and settlements fragile and imposes additional costs on new investment in pipelines and land-based factories, mines, and oil and gas fields, the further development of the inland arctic landmass becomes problematic. But the long-blocked alternative of development by sea now beckons. We have already looked at some of the implications of the Northern Sea Route in the preceding chapters on oil and gas, but in the second half of this chapter I focus more specifically on the politics behind it, and its potential consequences for the development of the Far North itself. At present, the two dominant trends in the North—active development along the coast and relative stagnation inland—are proceeding independently, as Moscow focuses on the first and neglects the second.

This two-track policy has come under criticism from Russian arctic specialists—many of them based in Siberia themselves—who point out that it aggravates the worst of the hydrocarbon model. The development of the oil and gas industry along the Arctic coast is dependent on imported technology, and ultimately increases the vulnerability of the region to global export trends. They advocate instead a policy of "connectedness"—integrated development based on local industry in existing Siberian centers such as Novosibirsk and Irkutsk. But it is not clear that they have realistic solutions, particularly to the lack of north–south transportation infrastructure. So far their voices are going largely unheard.[23]

The Northern Sea Route: A Beneficiary of Global Warming

The Northern Sea Route (NSR) extends more than 5,500 kilometers from Novaya Zemlya to the Bering Strait.[24] It is still icebound for about eight months of the year, and until recently it has been inaccessible to navigation during that period for all seaborne traffic unless escorted by powerful icebreakers. In Soviet times the government began building a fleet of nuclear icebreakers and promoted the NSR as a prospective commercial route linking Europe and Asia via the Arctic Ocean. But with the fall of the Soviet Union, the construction of new nuclear icebreakers stopped and shipping activity virtually ceased. The NSR remained a dead letter. Traffic was confined to the warmer waters of the Barents Sea, with the sole exception of Norilsk Nickel, which operated its own icebreaking cargo fleet to support its mining activities in northern Krasnoyarsk Kray.[25] There was no through traffic to East Asia.

But interest in the NSR has revived strongly in the new century, as the scale of global warming has become more and more evident. With the launch of LNG production from the Yamal Peninsula in late 2017, traffic to East Asia began growing steadily, although as yet only in the summer and with the help of icebreakers. In July 2018, history was made when an icebreaking LNG tanker crossed the NSR to Asia with an icebreaker escort.[26]

The prospect of an ice-free passage between Europe and Asia has revived the Soviet vision, espoused by Mikhail Gorbachev in 1989, that the NSR could open up a major new commercial sea route, offering a significantly shorter transit time from Europe to East Asia than the roundabout Suez Canal.[27] In addition, as we have noted, an open seaway would accelerate the economic development of the Far North along the Arctic coastline, chiefly through increased production and export of oil and gas and metals. The Russian shipbuilding industry would also benefit. A further advantage, of interest to the Russian military, is that it would lead to a reinforced Russian security presence in the Arctic Ocean through the construction of new ports and bases. Thus, insofar as climate change brings benefits to Russia, the NSR is at the top of the list.[28]

Putin's interest in the NSR is of long standing. In April 2000, shortly after his election to his first term as president, he delivered a major speech in Murmansk on marine transportation to a gathering of senior officials, in which he called for 10 million metric tons (MMt) of shipping annually via the NSR. He returned to the subject in another major address in Murmansk in 2007, this one held on the nuclear icebreaker *50 let Pobedy* (*50 Years of Victory*). Even at this early date, Putin invoked global warming as a major opportunity for expanding the route to the east.[29]

In his annual address to the Russian legislature in 2018, Putin raised the priority of the NSR further, setting a goal of 80 MMt of annual traffic by 2024. Now the NSR has become part of a vast program of "national projects," Putin's plan for the economic and social redevelopment of the country. The NSR is one of six federal projects in the overall program to develop the east–west and north–south transportation corridors. But in this collection, one point stands out: the priority of the Arctic. As Putin observed at the 2019 Arctic Forum, 10 percent of Russian investment is currently going into the Arctic, and the share is likely to grow.[30] Much of it touches, directly or indirectly, on the development of the NSR.[31]

The Northern Sea Route is thus being touted as the key to the future of the Arctic, and for the Kremlin it represents the chief benefits of climate change to Russia. The benefits would be both direct and indirect.

The direct transit route between Europe and western Russia to the rapidly growing markets of Asia would be shorter by two to three weeks than the passage through the Suez Canal. For the sake of comparison, currently about 20,000 vessels pass through the Suez Canal each year, earning Egypt an average of $5 billion annually.[32] But the real prize for Russia is the indirect returns if the sea route stimulates the development of the entire Arctic coastline, particularly for export of oil and LNG. In addition, there would be nonquantifiable benefits, such as increased capacity for military power projection. Accordingly, the focus of Kremlin policy has shifted in recent years from transit to the development of the hydrocarbon resources for export, primarily to Asia.[33]

Whether that vision is ultimately realized, and at what rate, will depend on several things. The first is the future of Asian demand for Russian raw materials. As I have argued in the preceding chapters, revenues from Russian oil, gas, and coal in Asia are likely to grow for the next decade, but peak thereafter, and decline by mid-century. On the other side of the ledger will be the considerable cost of developing the NSR, as well as the infrastructure to support the next generation of oil and gas along the coast. The challenges are formidable, beginning with the cost of shipping.[34]

Shipping Is the Key Constraint

The ultimate attractiveness of the NSR depends on the rate at which the ice melts and the eastward route clears. The westward portion, from the Ob estuary to Murmansk, is already in heavy use, notably to move exports of oil to Europe. But the main target, in Russian eyes, is the eastern route, which will connect Russia's LNG exports to the Asian market. Here the crucial question is transportation. No matter how fast the ice melts, the eastern route will still require ice-worthy shipping for the foreseeable future. Putin's target of 80 MMt per year of traffic crossing the NSR by 2025 has since been raised further: the recently approved Strategy for the Arctic calls for shipments of 120 MMt per year by 2030 and 160 MMt per year by 2035. But there is not yet enough ice-capable shipping available to meet this target.[35]

Shipping is thus the single most important issue in the development of the NSR, and this has given an increasingly powerful role to Rosatom, the state's nuclear conglomerate. We have already met Rosatom in several of its activities, notably in nuclear power and renewables. But Rosatom is involved in shipping as well: it is the parent company of Rosatomflot, the owner of the Russian civilian nuclear fleet. In 2018, Rosatom was named the overall operator of the NSR. This decision came after several years of mounting dissatisfaction at the top of the government with perceived "foot-dragging" by the Ministry of Transportation, which had previously been named to lead the NSR program. Rosatom had the support of key players in the presidential administration, and has now emerged as a lead player.

The logic of Rosatom's growing role is that for the next two decades, despite the melting ice, most ships plying the NSR will still require icebreakers. Rosatom's answer, not surprisingly, is nuclear power. But at the beginning of the 2010s Rosatom's nuclear fleet consisted of only five aging Soviet-era icebreakers; no new nuclear vessels had been constructed in the preceding thirty years. Beginning in 2012, Rosatomflot ordered the first of a new series of nuclear icebreakers from the Baltic Shipyard (Baltiysky Zavod) in Saint Petersburg, to provide escort services along the eastern route.

Reviving the construction of nuclear icebreakers after the long post-Soviet hiatus has had problems; as a case in point, the completion of the *50 Let Pobedy* took eighteen years. Historically the Baltic Shipyard was part of the Soviet military-industrial complex and specialized in the building of warships. After the fall of the Soviet Union, the Baltic Shipyard mostly lay dormant. Now, with the revival of military spending under Putin, the Baltic Shipyard is busy once more with military contracts, and civilian business is again taking a backseat.[36] This is apparently a problem with all of the major shipyards.

Putin has taken notice. In a September 2019 meeting with the Military-Industrial Commission, Putin scolded a group of military-industrial officials for their inadequate efforts in civilian projects. He particularly singled out delays in the plan for "shipyards that traditionally produced warships" to build nuclear and diesel icebreakers for the NSR.[37]

Despite delays, though, Rosatom's program has forged ahead. Its first three new nuclear icebreakers are scheduled to be completed by 2022,[38] and Rosatomflot has signed another contract with the Baltic Shipyard for two more icebreakers of the same series. By the early 2030s, Rosatomflot's nuclear-powered fleet will have grown to eight vessels that will be available for escorting traffic through the NSR. But Rosatom's ambitions do not stop there. In April 2020 it commissioned the first of yet another series of nuclear icebreakers, the *Lider* ("leader" in Russian), 209 meters long and equipped with two reactors generating 315 MW of power. The *Lider* will be far and away the most powerful vessel on the NSR, capable of plowing through ice 2 meters thick at 12 knots, and even ice 4 meters thick at 2 knots.[39] It is supposed to be launched in 2027. However, there are still questions about the *Lider*. The new yard in which it is being built, the Zvezda (a special project under the direct authority of Rosneft chairman Igor Sechin), has a history of problems, and it has yet to demonstrate that it can build even conventional vessels on budget and on schedule, let alone a major next-generation nuclear vessel.

A second question, more fundamental, is whether the *Lider,* and indeed Rosatom's entire nuclear fleet, will actually be needed. An emerging rival program is an alliance of Novatek with the shipping company Sovkomflot. Sovkomflot has already demonstrated that it can ship oil eastward without Rosatomflot's services; and the two companies are exploring the concept of non-nuclear ice-capable LNG tankers powered by the ship's own LNG (so-called *gazovozy*). Novatek's CEO, Leonid Mikhelson, in particular, argues that the *Lider* is not needed. The combination of warming seas, more-capable ice-class LNG tankers, and regular shipments that keep the sea lanes open, all indicate that icebreaking escort for each voyage may not be necessary much longer.

Conclusions

The key point of this chapter is that the main response of the Russian political and business leadership to climate change in the Far North and East has been the development of the Arctic coastline, primarily to increase exports of hydrocarbons. Regional adaptation of the inland to melting permafrost, as in the case of Yakutsk and other northern and eastern cities, takes a distant second place.

These policies suggest that, in practical terms, the region is envisioned as two different Arctics. The first, driven primarily by hydrocarbon development along the coast and increased exports through the NSR, is likely to grow so long as there is global demand to support it, despite the mounting costs imposed by melting permafrost and coastal erosion. In contrast, the second Arctic, consisting of smaller towns and mining centers inland, will continue to be neglected in favor of coastal development; and much of its native population will migrate toward the handful of established urban centers such as Yakutsk. Within the second Arctic, there will be little investment in new infrastructure and communications. Despite warmer temperatures, agriculture will be held back by poor soils, diseases, and lack of manpower, compounded by local resistance to foreign migrant labor.[40]

Critics of Russia's hydrocarbon model denounce this approach as a continuation of Russia's historic misallocation of resources, especially in view of the likelihood (discussed in Chapters 2 and 3) that the role and revenues of fossil fuel exports will decline as the century proceeds. They call for increased diversification and more integrated development within the Siberian region. Yet where the Far North and East are concerned, it is hard to see what such diversification would consist of, given the growing costs of investment in the inland and the uncertain long-range returns. Thus, the high-priority development of gas and oil, concentrated on the seacoast and oriented toward exports via the NSR—the "first Arctic"—is a rational policy for the present. But beginning in another decade, as oil demand peaks and the growth of

LNG demand begins to slow, the costs of neglecting the Arctic inland will loom increasingly large.

Russia's policy toward the Arctic coastline and the Northern Sea Route has gone through constant revisions and exemptions as its priority has risen, particularly as the result of the emergence of LNG as a major factor (see Chapter 3). As recently as 2013, the focus was on the NSR's potential for international transit, and the main goal of policy was to increase the attractiveness of the route for foreign shippers. But beginning in 2014, the scene changed. As Russia's relations with the West cooled, its focus shifted from transit to security, and the military became more involved. At the same time, the growing prospects of LNG exports from the Yamal Peninsula introduced new players, bringing new conflicts. As we have seen, Novatek's demands for latitude to use foreign-made and foreign-flagged shipping, in order to increase LNG exports as quickly as possible, have been opposed by a coalition of industrial and political interests, which insist on using Russian vessels. At the same time, the increasing priority and commercial attractiveness of the NSR have drawn in an ambitious new player, Rosatom, which is promoting a new generation of nuclear-powered icebreakers and is using its leverage over transportation to bid for control of the entire NSR—against the vigorous opposition of Novatek and Russia's largest shipper, Sovkomflot.

Faced with this swirling mix of competing players and motives, the Kremlin has temporized by granting something to each one. It has granted a series of exemptions to Novatek while at the same time tightening the rules for import substitution, and it has supported the ambitions of Rosatom while refraining for the time being from an all-out militarization of the NSR. Meanwhile, in the background, the melting of the Arctic ice is accelerating, increasing the ambitions of all the players, and deepening the Kremlin's conundrum over how to reconcile them all.[41] Nevertheless, what is clear is the Kremlin's overall commitment to the "coastal Arctic" as its main response to climate change, as opposed to the "inland Arctic." That will not change.

9

METALS

Oleg Deripaska, the founder of Russia's monopoly aluminum company, is the ultimate survivor. Like many of the oligarchs who dominate Russian industry today, he is not a specialist in his field by background. He began trading in aluminum while still a student at the Physics Faculty of Moscow State University in the early 1990s. By that time the aluminum industry was well on its way to chaos. In Soviet times the industry had relied on imports of bauxite and alumina, the basic material from which aluminum is produced, which was then turned into aluminum using the abundant hydropower developed in the 1960s in Eastern Siberia. In the 1980s this supply chain was disrupted by the Gorbachev reforms and the end of the Soviet foreign-trade monopoly. Foreign traders quickly moved in and made rich profits, importing alumina into the country, refining it into aluminum for a tolling fee, and re-exporting the pure metal.

Thanks to his early ties with traders, the young Deripaska, a year after graduation, was installed as general director of one of the largest Siberian producers to represent the traders' interests. But within two years he turned the tables on his patrons: he diluted their shareholdings, canceled their trading contracts, and drove them out of the company. Two years more, and Deripaska was already at the head of a network of producers previously controlled by the traders. From there his path only led upward. By the mid-2000s he had assembled a company, Rusal, that quickly became the dominant aluminum producer in Russia and the second-largest in the world. Even by the standards of the "Wild East" Russia of the day, Deripaska's rise was perhaps the most remarkable of all.[1]

Today, under its parent holding company En+, Rusal has developed into a powerful international conglomerate that controls the entire production chain from the extraction of bauxite and its processing into alumina, to the smelting of aluminum metal and its use in the manufacture of a wide range of products, including military aviation. (Indeed, in Soviet times aluminum was treated as a strategic industry.)

However, it is Deripaska's next steps, and his present role in the politics of climate change, that are of interest for us. Rusal's overseas expansion made it a major emitter of greenhouse gases, and an increasingly visible target for governments and financial groups as the global campaign against climate change gained momentum. Rusal quickly got the message, and it began greening its international image and operations. Inside Russia, Deripaska became an energetic campaigner for a more active policy to control emissions, against the resistance of most of Russian industry. In 2019 Deripaska, faced with US sanctions, divested himself of direct ownership of Rusal, but he remains a strong voice in the politics of climate change.

This chapter tells the story of the metals industry, both as earners of export income and as players in the emerging politics of climate change. The metals sector is difficult to capture in a single chapter. Each group of metals is produced by separate companies with their own histories, their own technologies and infrastructures, their own economics—and their own carbon footprints. Nevertheless, the various Russian metals producers do share certain key features. First, the companies in this sector are private, owned by major oligarchs, and largely independent of direct state subsidies.[2] Second, the carbon footprint of the metals sector as a group is relatively minor, compared to the energy sector. Third, the metals companies have expanded vigorously into export markets and overseas production, and as such they are some of Russia's most global players. But as producers of tradable commodities they are all subject to the vagaries of world markets and prices, which make their earnings unpredictable.[3] Fourth, metals are embodied in the manufacture of other export goods, such as airplanes and weapons, and in those embodied forms they add further to Russia's export revenues, but also to the carbon footprint of those items. Finally, Russian metals exports stand to be seriously affected by the imposition of carbon border taxes by the European Union (EU) (so-called CBAM, discussed in Chapter 1), unless they can cut their carbon footprint to satisfy the EU's requirements.

The Russian metals sector is a far smaller source of CO_2 emissions than the power sector. In 2016, out of a Russian total of 2.6 billion tons of CO_2 emissions, industry accounted for nearly 9 percent, of which metallurgy emitted about half. To put this in perspective, the heat and power sector typically accounts for over 82 percent of Russia's CO_2 emissions from industry. In short, CO_2 emissions in Russia are primarily a power-generation issue, not a metals issue.[4]

The metals sector is also a significant—but again, not overwhelming—exporter, accounting for 6.7 percent ($28.6 billion out of a total of $424.6 billion) of Russia's export revenues in 2019. This is dwarfed by the export earnings of the oil and gas sector, which, as we have seen, totaled $237.7 billion in the same year, or 56 percent of Russia's total export income. Oil and gas make the weather in Russia, but not metals.

Thus, the metals sector would appear to be a relatively minor part of our equation in this book, were it not for the fact that a handful of companies in the sector, as well as some individuals such as Deripaska, have played surprisingly visible roles in Russia's climate politics, as advocates of greater climate regulation and controls. This contrasts with the behavior of the coal industry, which has resisted the climate-change story and opposed limits on its emissions.[5]

The different behavior of the metals sector appears to be due to the fact that the metals companies as a group have large direct stakes outside Russia, having expanded both vertically (buying foreign providers along their own value chain) and horizontally (acquiring foreign companies in the same final business as themselves). In this respect Rusal has played the lead role, and the other metals producers less. The

Russian metals producers were able to do this because they had export income from the beginning, whereas the coal industry was initially largely focused on the domestic market and did not develop significant export revenues until later on. As a result of their foreign holdings, the Russian metals companies are more visible abroad and more subject to outside pressures over climate, and they have a correspondingly greater incentive to develop green credentials. In contrast, the coal sector is more inwardly focused, particularly because of its large role as an employer in coal-producing regions.

Climate change outside Russia is likely to affect metals exports differently from other Russian exports. Above all, unlike oil and coal, metals will not face disappearing markets. But different metals will be affected differently. Aluminum displaces steel, where lighter weights are at a premium. Demand for key nonferrous metals such as nickel and copper is booming, driven by the growing demand for batteries of all kinds. Demand for "exotic" metals such as platinum and cobalt is rising strongly for the same reason. These trends are likely to continue in the future. Overall, however, metals exports are likely to lag behind global GDP growth, owing to the declining share of industry in global GDP and the historical trend toward declining metals-intensity in the global economy.

This would mark a return to the historical norm, after the great commodities supercycle of the last twenty years, which was led above all by the metals sector, and especially by demand from China. The boom in Chinese demand for metals was astounding. Between 1997 and 2017, according to the World Bank, China's share of global metals consumption rose from 10 percent to 50 percent. China accounted for four-fifths of the increase in global metals demand in those years.[6] But China's future demand for metals is a wild card. On the one hand, any slowdown in China's economic growth would have a disproportionate impact on metals in particular. On the other hand, the widespread adoption of electric vehicles and the spread of solar power and wind, for example, could drive continued growth in Chinese demand for specific metals such as copper, even if China's overall growth rate slows.

China plays a dual role in steel. China is the world's largest exporter, well ahead of Russia, which holds second place. At the same time, it is also the world's eighth-largest importer, mainly from neighboring Asian producers but not from Russia. Thus on balance it is more of a competitor for Russian steel than a customer. If the pace of infrastructural investment in China slows and excess Chinese steel production is reoriented toward the world market, Russian steel exports would come under severe pressure.

In the longer term, displacement of metals by synthetic materials, and widespread adoption of additive technologies (such as 3-D printing) in place of traditional manufacturing, could lead to an erosion of demand for metals, especially for steel in the automotive sector.[7] Yet the overall impression is that these new technologies will be slow to take significant market share by mid-century. The contribution of metals exports to Russia's growth and revenues may gradually diminish as a result of climate change and technology, but without sudden crises.

That said, each metal is its own separate story. In this chapter we look at the contrasting cases of aluminum, nickel, and ferrous metals.

Aluminum

From the standpoint of the politics of climate change, the most interesting metal is aluminum. The sole Russian producing company in the sector today is Rusal. Under Deripaska, as we have seen, Rusal became a global giant. Aluminum is Russia's second-largest metal export, with $4.6 billion in revenues in 2019, compared to $18.1 billion for steel and iron.[8] Rusal controls more than forty plants in thirteen countries and produces 7 percent of the world's aluminum.[9] As mentioned, Russia depends on other countries for imports of bauxite and alumina, the raw materials for aluminum, and Rusal controls major producers of bauxite around the world, notably in Guinea, which has the world's largest bauxite reserves.

Worldwide, aluminum production is a powerful source of greenhouse gas emissions, accounting for over 1.1 billion tons of CO_2 equivalent in 2019, about double the level of 2005.[10] All told, aluminum production emits 11.7 tons of carbon for every ton of smelted metal, while steel emits only 2 tons. The precise amount varies widely, depending on each producer's efficiency, the age of its equipment, the source of its energy, and the quality of its raw material, bauxite. As one insider remarks, "The emission profile of aluminum production is huge and very disparate. It can be produced with four or five tons of CO_2 or 20—and more than half the industry does it with 20."

Rusal is somewhere in the middle. Most of the carbon footprint of aluminum comes from the processing of bauxite and alumina. Rusal imports most of these, so its carbon footprint is mostly located outside of Russia. Inside Russia, because Rusal's aluminum ore is smelted into primary metal using hydropower, it has only a minor carbon footprint at home. (Aluminum is sometimes described as solid electricity.) The main pollutants at this stage are fluorides and perfluorocarbons, not CO_2.[11]

This puts Rusal in a different category from other Russian metals producers because its visibility in climate-change politics mostly arises outside of Russia. This helps to explain the proactive climate policy pursued by Rusal. Rusal advertises its green profile, and it has come out in support of steep carbon taxes on energy-intensive Russian exports such as coal and ferrous metals.[12] En+ has pushed the London-based Metals Exchange (LME) to force aluminum producers to divulge their CO_2 footprint.[13] As early as 2008, Rusal joined the UN-sponsored Global Compact "Caring for Climate: Business Leaders Platform for Action," which promotes business initiatives on climate.[14] (Rusal was one of only two Russian companies to do so.) In 2015 Rusal joined the international project on releasing information about CO_2 emissions. Deripaska supports the introduction of a global emissions tax.[15]

Inside Russia, Deripaska, has played a visible role in Russia's climate-change debates, as an advocate for stricter legislation. As chairman of the RSPP Environment Committee, Deripaska pushed actively for Russian legislation to adopt the "best available technology" principle, under which industries must modernize their production processes to achieve the best world environmental standards.[16] In 2019 Rusal announced that it had cut its emissions per ton by 7.5 percent over five years, and it was aiming for a similar cut by 2025, although part of that will apparently be achieved when Rusal divests its remaining coal-fueled production.

The Chinese market is critical for Russian aluminum. Even though the headlong growth of the Chinese economy has slowed, Chinese demand for aluminum is likely to remain strong. The Chinese automotive industry continues to expand, while simultaneously shifting to lighter-weight vehicles, and consequently its consumption of aluminum could well double by mid-century.[17] However, China produces half of the world's aluminum on its own, ten times more than Russia. By 2019 China already had unused capacity equal to twice Russia's production.[18] Thus, Russia is at best a swing supplier to the Chinese aluminum market, and this vulnerability will worsen as time goes by. Russia's one comparative advantage is that Chinese smelters are still largely fueled by coal. Hence, Russia's exports to China will depend partly on how quickly China moves to other fuels (see Chapter 4).

Nickel

The fuel spill at Norilsk in May 2020, recounted in Chapter 8 on the Arctic, made Nornickel a national symbol of ecological mismanagement in the Far North. The episode was ironic, in the sense that Nornickel was already investing heavily to control its main pollution problem, the production of sulfur dioxide, and to improve its governance procedures. In particular, in 2016 Nornickel shut down its oldest smelter at Norilsk, and in 2019 it began closing its smelting plant in the Kola Peninsula, located close to the Norwegian border. Putin had long shown interest in Nornickel's pollution reduction program, and in December 2018—eighteen months before the spill at Norilsk—its principal owner, Vladimir Potanin, met with Putin to report on its progress. At the time Potanin claimed that the company had allocated 150 billion rubles (then about $2.5 billion) to its overall environmental program, mainly to cut emissions of sulfur dioxide.[19]

But the spill at Norilsk illustrated the difference between a classic "environmental" problem—air pollution and acid rain—and climate change, which threatens to destabilize infrastructure throughout the Far North. In the long run, adapting to climate change would cost far more than conventional environmental protection (see Chapter 8). But fortunately for Potanin, he has not been given that wider mission. Unlike the aluminum and steel industries, Nornickel does not produce significant amounts of CO_2. Consequently, if it can stay out of the limelight in northern Russia, the nickel industry might remain relatively unaffected by the domestic politics of climate change.

Moreover, as Russia's leading producer of copper, palladium, and platinum, and lately rhodium as well as other rare-earth metals, Nornickel could play a growing role in Russian metals exports. Until now, Russia has not made a major effort to develop exports of these metals; instead, Russia has relied on imports for its own needs. However, global demand for these "new" metals is likely to grow strongly in coming years and could be a source of additional export revenue.[20]

Steel and Ferrous Metallurgy

Within the Russian metals group, ferrous metallurgy is the largest emitter of CO_2, accounting for about half of the total from the sector.[21] Most ferrous metals in Russia are produced with coal; indeed, in post-Soviet times several large steel companies, such as Evraz and Mechel, have acquired large coal assets and have become coal exporters themselves.

Traditionally the main markets for Russian ferrous metals have been Europe in the west and China in the east. Because of high transportation costs, more-distant locations have not played a major role. In 2019 iron ore and steel exports earned $18 billion, or about two-thirds of all metals exports.[22]

Russia produces both iron ore and steel, but while most Russian iron ore is exported, Russian steel has been largely consumed inside Russia. Exports helped to preserve the steel industry in the 1990s, but major steel companies such as Magnitogorsk (Russia's second-largest steel producer after NLMK) shifted back toward the domestic market in the 2000s, and Russian companies now account for most of the steel industry's sales.[23] The steel industry is therefore less sensitive to fluctuations in world demand and prices than other Russian metals; but as the Russian economy has slowed, domestic demand for steel has declined as well, increasing pressure to increase steel exports.[24]

Although most other commodity exporters have been consolidated around a single dominant company, the ferrous metals sector remains divided among numerous producers, which compete vigorously with one another as well as with foreign exporters.[25]

Russian steel producers have invested heavily in modernizing their operations over the past quarter-century, putting over 2.6 trillion rubles (currently about $35 billion) into new processes and plants, although according to one leading Russian expert, they still lag behind the technological level of Western Europe.[26] Severstal, in particular, has drawn international attention for its automation and digitalization of its main production center in Cherepovets near Moscow.[27]

Little of this effort, however, has been devoted to cutting the sector's carbon footprint. Aleksey Mordashov, the main owner of Severstal, stands in interesting contrast to Deripaska. Mordashov has been a highly visible promoter of the "S" (social) and "G" (governmental) of ESG, but much less of the "E"—the environmental aspect of sustainability. For Mordashov, "climate" means mainly investment climate. At the 2018 annual meeting of the industrial lobby group RSPP, which was attended by Putin, Mordashov pointed to Russia's lag in productivity behind US industry, and called on Putin to reduce needless regulation and red tape.[28]

Yet the steel industry too has felt the growing pressure from international investors to improve its green image, and the leading companies have begun to issue regular reports on their emissions. Magnitogorsk (MMK) issued its first sustainability report in 2019. MMK and other Russian companies have made use of the Global Reporting Initiative (GRI), an international nonprofit that helps companies develop reports on a wide range of ESG issues, including greenhouse gas emissions.[29] Recently a number of Russian metals producers have taken advantage of "green finance" loans to implement best ecological technologies.[30]

The greatest climate-related threat facing the Russian metals exporters is the possibility that the European Union, Russia's largest export market for metals and commodities of all kinds, will levy carbon export taxes to penalize exporters with larger carbon footprints than those allowed to European producers.[31] Because Russian iron and steel are largely produced with coal, ferrous exports to Europe would be highly vulnerable. The European Union already imposes high tariffs and antidumping penalties on Russian steel exports, the Russians complain, although so far Russia's low costs and the weak ruble have enabled the Russians to maintain their market share. But that may now be under threat.

Conclusion: The Long-Term Outlook for Russian Metals

Climate change will impact Russian metals exports in two contrary directions. Iron and steel exports will be particularly vulnerable because they are still produced primarily with coal, and thus may be subject to European carbon export taxes in some form. The prize will go to the most modern Russian producers, and particularly those that use gas. But this may not be enough; the next generation of innovation in the European steel industry will be carbon-free steel, which would require a major overhaul of Russian steel to stay competitive.[32]

Aluminum and the nickel group, in contrast, are likely to benefit from increased global demand to supply electric vehicles, consumer electronics such as smart phones, and renewables. Export prices and revenues from this group are likely to increase.

These two opposite trends will broadly cancel one another out. Ferrous metals currently account for two-thirds of Russia's present metals exports; but that share is likely to go down. Aluminum, nickel, and palladium group metals, which make up most of the remainder, are likely to increase. The sum of these, however, is not likely to amount to much more than the roughly $30 billion (in 2018 dollars) that the metals sector earns today.

The more interesting question ahead is how the politics of climate change outside Russia, and especially in Europe and China, will affect climate-change politics inside Russia. Already the prospect of carbon export taxes is polarizing the climate-change debate, as Russian environmentalists and reformers use the threat from Europe as an argument to press for sterner controls on carbon emissions, especially carbon pricing. So far, Russian industry, led by the coal sector, has managed to hold the line against these demands. But over time some form of carbon pricing on metals seems likely.

Thus, the role of the Russian metals sector is important to watch as an illustration of how the growing pressures for decarbonization outside of Russia are transmitted into the country, driving domestic policy toward a more vigorous role in controlling carbon emissions. As we have seen, the direct contribution of the Russian metals sector to CO_2 emissions is relatively minor, but its impact on the politics of climate change will be increasingly important. But these are still early days.

CONCLUSION

The Reckoning Ahead

By 2050, climate change will have become humanity's most urgent concern. It will also be the dominant issue in world affairs. Climate change will have consequences for every country in the world, but for Russia's population, its economy, and its political system, the changes that come from climate change will increasingly shape Russia's condition at home and its position in the world.

Compounding the impact of climate change on Russia will be a major political event—the end of the Putin era. By the mid-2030s, if not sooner, Vladimir Putin will have left the political stage, closing a reign that by that time will have lasted longer than Stalin's. This book is not the place to speculate on whether the coming transition will be smooth or not, but one thing is certain: Russia under Putin will have expended much of the lowest-cost oil and gas resources with which it entered the century, while many of the fundamental liabilities inherited from the Soviet era will still remain unresolved. In the age of climate change, Russia's extractive export model, which has prevailed for the past thirty years, will be less and less sustainable. And simultaneously, the country's options to deal with the consequences will be increasingly limited.

In this book I have presented five major propositions about the coming effects of climate change on Russia over the decades to 2050. The first three concern the external consequences for Russia's economy; the other two deal with the direct impact of climate change on Russia's territory itself. The key argument is that climate-related trends outside Russia will affect Russia far more seriously than the direct, internal impacts, at least until the second half of the century.

Beginning in the 2030s, the external effects on Russia will be severe. Climate politics, technological change, and energy transition outside Russia will cause a sharp drop in Russia's income from the export of fossil fuels—mainly oil—which are the lifeblood of the economy. The resulting massive loss of wealth will affect Russian citizens at all levels but will especially impact the state, depriving it of the means to reform the economy or to address the increasingly serious direct effects of climate change on people's lives, as these worsen after mid-century. This in turn will have major consequences for Russia's position in the global economy, and its standing as a great power.

Some may argue, especially in Russia, that energy transition is not the same thing as climate change, and that in some respects climate change could even be beneficial for Russia. I argue that such a view is profoundly mistaken. The internal impact of climate change will be primarily negative. Any internal benefits it may bring to Russia will be overwhelmed by the impact on Russia of the energy transition in the rest of the world.

External Consequences: Russia's Declining Export Revenues

The main impact of climate change to 2050 will be on Russia's export revenues and its economy. The findings of the preceding chapters can be summed up in the first three propositions:

First, as early as the 2030s, the world's transition away from oil will be well under way, coal will be in irreversible decline, and even gas will be nearing its peak. Russia will have little control over these global trends, but together they will amount to a game changer for its economy. Russian revenues from fossil fuels will be in steady and irreversible retreat.

Second, Russia has no substitutes for fossil fuels as money earners. Natural gas and liquefied natural gas (LNG) will make up only a small fraction of the shortfall from oil. Coal exports will fade as the world's demand for coal declines. Russia's ambitions of becoming an exporter of high technology will not be realized, because of chronic weaknesses in technological innovation and commercial realization. The only significant exceptions will be nuclear power and weapons, but these will not fill the gap.

Third, as a result of the decline in export revenues, the state's resources will be lastingly diminished. A growing share of the declining income will need to be reinvested in the natural-resource sectors themselves, especially oil. This will cut into the residual available for redistribution to the rest of the economy for infrastructural investment, subsidies, defense, security, and social welfare. Growth rates are likely to weaken, and living standards to stagnate. Political competition for the state's remaining resources will intensify.

These first three trends will be driven primarily by forces outside Russia, and especially by the speed of the energy transition. All global projections for climate change–related outcomes share the same basic conception of the energy transition: it is a contest between the growing forces for change (both positive and negative) and the inertia of established structures and systems. In this contest Russia is not a force for change; instead, it is part of the inertia. But its actions will have little influence over the global course and outcome of the contest, compared to those of the other major players—the United States and Europe, and above all China and India. Russia will be a taker of changes arising from outside, not a maker.

The external consequences of climate change for Russia are likely to unfold in two phases. During the first, extending through the 2020s, the "slow transition" path will govern. Oil and gas consumption will continue to increase, while coal will retain a lease on life. Commodities prices may even continue to rise in that decade, and there will be occasional spikes in response to passing events. During this period, the present hydrocarbon model of Russian policy will remain temporarily viable.

But by the early 2030s, a second phase will take over. As the effects of climate change become more pronounced around the world, the fast-transition path will increasingly prevail. Oil demand will peak sometime during this decade and then gradually decline. While global gas demand will continue to grow, renewables will increasingly push coal out of the energy mix. Russia's export revenues from energy exports will begin an inexorable decline. By 2040, the fast-transition path will have become dominant, and from that point on the decline in fossil fuels will be rapid. Export income from other sources will not make up more than a small part of the lost revenue from fossil fuels.

The overall implication is dramatic: for the first time since the 1960s and 1970s, when the Soviet Union began exporting oil and gas on a large scale, Russia will be unable to fuel its growth by exporting energy, and it will lack the capital to invest in adaptation and modernization. We pause now to elaborate on these first three propositions, before going on to the remaining two, which concern climate change inside Russia.

Russia's Export Revenues and the Coming Budget Gap

The projection of future export revenues in the Introduction (Table 1) suggested that Russia's income from exports might decline by as much as 45 percent by 2050. However, this represents only gross revenues. From these, one must subtract the investment and operating costs of the producers, which will be higher than today's. The remainder will be the proverbial shrinking pie, which must then be divided between the producers and the state. In considering these numbers, one should bear in mind that these trends will continue beyond mid-century and that Russia's export income is likely to continue to deteriorate.

The point is not that Russia will lack natural resources, but that demand for them will weaken, owing to the spread of renewables, growing restrictions on the use of fossil fuels, and the electrification of an increasingly service-based world economy. Lastly, the pace of Chinese economic growth, which powered the explosion of commodity demand over the last generation, will slow. The great commodity super-cycle of those years will not be repeated. Russia's remaining hydrocarbon resources will be increasingly trapped between declining demand abroad and increasing cost at home.

What will be the impact on the state's ability to spend? Over the last decade the share of oil and gas export revenues in the federal budget has swung between one-quarter and one-half, depending mainly on variations in world hydrocarbon prices.[1] If one includes indirect export revenues, such as dividends paid to the state as shareholder, the share of oil and gas exports in the state's revenues is even larger. In contrast, the non-hydrocarbon portion of the federal revenues is consistently in deficit. In other words, it is only thanks to the oil and gas export revenues that the Russian government is able to balance its budget. These are critical numbers, given the key role of the state in supporting the investments of state-owned companies, as well as providing subsidies, funding state programs (including defense), paying civil servants, and maintaining the country's vast welfare system. Russia's aging population will require more spending for pensions and health care. Will the state's revenues be enough to cover these expenses?[2]

Over the past decade the Gaidar Institute in Moscow has run a series of projections of what it calls the budget gap, which is defined as the balance between the present value of the state's revenues against its expenditures.[3] The latest projection, based on the Ministry of Economic Development's long-term forecast for economic growth to 2036, predicts a chronic budget gap between 9.4 and 10.1 percent of GDP.[4] The two main causes are the aging population and what the Institute calls "the unresolved problem of replacing oil and gas revenues."

The latter is the crux of the problem. Russia's entire political and economic system rests on the formal and informal distribution of "rents"—payments, especially those from fossil fuels, many through multiple channels that bypass the formal budget. These include the direct social costs covered by the producers themselves (especially the coal industry), the hidden subsidy provided to gas consumers in the form of low domestic prices, and the cost of padded construction contracts awarded to favored interests, as well as the steady leakage of profits abroad by various "informal" channels. As the best reserves are produced and production costs go up, and as export revenues decline, the flow of available rents will decline throughout the system, with potentially destabilizing consequences, both economically and politically.

The Impact on Russian Territory

Beginning in the 2030s, the direct effects of climate change will also be increasingly felt on Russia's territory itself. These can be summed up in two final propositions:

First, the Russian Arctic will continue to warm faster than the rest of the globe. Permafrost, which takes up 70 percent of Russia's total land area, will melt at an accelerating rate, with increasingly severe economic and social consequences. Infrastructure, such as roads and buildings and pipelines, will be damaged. Construction will become more costly. Investment will be less profitable. Life will become harder for Indigenous populations, owing in particular to the spread of parasite-borne diseases and the growing difficulty of nomadic herding. Migration out of the Arctic will accelerate, except for the hydrocarbon production sites along the Arctic coast. The one positive result of global warming will be the development of the Northern Sea Route, as sea ice thins. This will facilitate the expansion of LNG exports to Asia. But the overall revenues from traffic through the Northern Sea Route will remain modest, as commercial and security restrictions deter European shippers.

One important consequence of the warming of the Arctic will be increased emissions of methane, released from melting permafrost into the atmosphere. Despite the growing publicity being given to this problem, there is yet little precise information available about it. Methane currently accounts for about 20 percent of total global greenhouse emissions, but two-thirds of it comes from agriculture and natural sources such as wetlands; the other third comes from the energy sector, much of it from coal mines and gas distribution systems. Methane is a serious addition to the problem of greenhouse gas emissions because it has twenty times the impact of CO_2, although methane disappears rapidly from the atmosphere. There is no way to eliminate this effect because methane is continuously generated within the earth's crust, together with decomposed organic matter within the frozen soil itself. Methane emissions from permafrost are still a relatively minor source compared to all others (on the order of 4 percent or less), but they will grow rapidly as permafrost melting accelerates. The increasing use of satellite-based detectors may soon make it possible to obtain a better picture.[5]

Second, in parallel with the problem of melting permafrost from the warming Arctic, more frequent droughts in Russia's southern regions will diminish grain yields and slow the growth of Russia's grain exports, which have been a major success story in recent years. It will be difficult for Russian agriculture to adapt by expanding to the north, despite generally warmer temperatures, because of the lack of fertile new soils. Consequently, although Russians will likely remain adequately supplied with food, grain exports may decline. Export income from agriculture, even in the best case, will not offset significantly the loss of revenue from fossil fuel exports.

In sum, climate change and the energy transition will bring to an end the era of abundant natural-resource revenues that has defined most of Vladimir Putin's time in office. Even now, high global commodities prices have not been sufficient to keep the Russian economy growing at more than a snail's pace over the past decade. As time goes on, the state will have fewer means available to cope with the consequences. This will be a major change, in view of the leading role that the state has always played throughout Russian history, and particularly in the last century, in both the Soviet and the post-Soviet eras. This in turn will lead to a number of further consequences. In what follows I consider four additional effects: popular reactions and possible political repercussions, the constraints on economic diversification and adaptation, the likely lack of progress in controlling emissions, and the consequences for Russia's behavior in international climate diplomacy. In closing we will look at the implications for Russia's place in tomorrow's global economy and its future position as a great power.

The Possible Political Impact of Climate Change

Paradoxically, the direct effects of climate change on most people's lives inside Russia will remain limited, at least in the first half of the century, because only 7 percent of Russians live in the Arctic and only 11 percent live in the southern agricultural region.[6] Most of the population is concentrated in cities located far from melting permafrost.[7] Consequently, climate change per se will not become a major domestic issue in the foreseeable future, as most Russians are likely to remain focused on traditional environmental issues that affect them directly, such as pollution or waste management. Extreme events, such as drought, heat waves, or forest fires, will periodically generate public emotions, but they are unlikely to become a sustained force for political action outside the regions directly affected, although the political impact of fires may increase as they continue to spread.

It is the indirect effects of climate change that will have the greatest political resonance, as people react to slower growth rates and declining living standards. Climate change could become an "ancillary" issue in domestic politics, especially via social media, if it becomes added to the total agenda of opposition politicians. So far, despite the efforts of nongovernmental organizations, this has not yet happened, as the opposition remains focused on issues like corruption and pensions.

This, at least, is the most likely picture for the coming decade. But in the 2030s, as today's Generation Z comes to maturity,[8] the political fallout from climate change may be considerably different. This is the first Russian generation to have grown up in the age of the internet, and it is strongly exposed to the increasingly anxious global conversation about climate change. Depending on when the post-Putin succession begins and how it proceeds, climate change could become a major political issue.

Limits on Economic Diversification

Taken as a whole, the history of the Putin era can be summed up as one of the dominant political power of the hydrocarbon lobby, backed by Putin's closest allies and underpinned by the security establishment. That feature has been widely condemned, both by Russia's domestic reformers and by foreign observers. Yet the Putin era deserves more credit than it usually gets. In each of the sectors considered in this book the Putin years have witnessed considerable achievement. After the collapse of the 1990s, each one was reconsolidated by a mix of state and private initiatives, each has been modernized and expanded, and new technologies have been adopted. In the oil industry, fracking and horizontal drilling are now the norm. In the gas industry, an entire new generation of gas supply—the Yamal Peninsula—has been developed, complete with an elaborate new pipeline export system. A new technology, LNG, has been developed from scratch. The coal industry has been reorganized around exports. The fact is, in a world of rapidly expanding energy demand, as the last generation has been, Russia's exploitation of its comparative advantage in tradable natural resources has been a rational policy. Some other sectors have been improved as well: Agriculture, as we have seen, has been rescued from its post-Stalinist doldrums, thanks to a combination of state support and private investment. These changes have been backed by effective monetary, fiscal, and budgetary reforms and policies,[9] which have put Russia in a strong financial position.[10]

But the revenues from this policy have also been massively misused, especially those from oil and gas. A unique opportunity to use the legacy rents inherited from the Soviet period—to skim the Soviet cream, so to speak—and to put them to work for the renewal of Russia, has been lost. Much of the legacy oil and gas rent has been dissipated without benefit to Russia, in the form of exported capital, via both licit and illicit channels, and through wasteful energy consumption at home. To be sure, not all has been wasted. Thanks to Putin's conservative financial policies—associated above all with the name of his long-time finance minister, Aleksei Kudrin—the Russian state has accumulated a nest egg that on the eve of the COVID-19 pandemic had reached nearly $800 billion, the equivalent of over three years of oil and gas exports. But Russia could have done so much more.

Putin's policy has deepened Russia's dependence on natural-resource exports, and the longer it continues, the more vulnerable Russia will become, as the effects of climate change, both external and internal, gather strength. Elite conversation in Russia today is just beginning to recognize this problem. There is much talk about the need for economic diversification, for a breakout into next-generation technologies, preferably exportable, on the model of successful developing economies. One well-known economist, Iakov Mirkin, a professor at Moscow's influential Institute of International Economic Relations and a specialist on international economic development, calls for an aggressive "industrial policy," not the accumulation of excessive reserves, but "surgery," as he puts it, instead of "ambulatory medicine."[11]

But one must question how realistic such views are, given the declining resources that will be available to the economy and the state. Russia's presently available nest egg, the product of twenty years of financial conservatism, will be quickly exhausted, and it will not be renewed. One of the main likely consequences of climate change for Russia is that an all-out policy of economic diversification will be much more difficult. As Russia's export revenues decline, and as the political succession approaches, there will be growing conflict between interests defending the long-dominant extractive model and those calling for economic diversification and increased spending on social and human capital. That is not a new debate—it has been a central issue ever since Putin was first elected in 2000. But it will grow more intense as the 2030s approach, and climate-change issues will become part of it.

This is especially true when one considers that Putin's successors will simultaneously face continuing challenges arising out of Russia's Soviet past.

The Continuing Weight of the Past

Thirty years after the end of the Soviet Union, Russia continues to pay the price for the damage done by the Soviet model of industrialization. Much of Russian industry to this day still consists of inefficient giants churning out obsolete goods that are uncompetitive on world markets, at a high cost in energy and raw materials. To the extent that investment has taken place, it has only deepened this pattern.[12]

The Soviet economic system has also left its mark deep in the layout of Russian cities. Russia has too many large cities located in remote places in the north and east, where they bear high costs. Paradoxically, climate change will only make this problem worse, as melting permafrost will impose even higher costs for adaptation of buildings and roads.

But the Soviet legacy does not stop there. Many Russian cities to this day are "monocities," developed by a single ministry around a single activity, whether military or space or natural resources. Large industrial units focused on just one part of the production chain, without any concern for subsequent processing (that was some other ministry's job). Entire cities were built around a single mineral, such as Norilsk and nickel, or Vorkuta and coal. Some of these, such as Norilsk, make economic sense today; many don't. Russia has over 300 "monotowns," which are home to over 14 million people, one-tenth of Russia's population, and account for 25 percent of the urban population and 40 percent of GDP.[13] Yet abandoning such "zombie cities" is difficult because of the lack of housing and job opportunities elsewhere. They survive on diminished state subsidies, simply because they are there.[14]

Undoing this legacy will require massive investment in industrial modernization and urban reconstruction and development, at a time when Russia will have less wealth to devote to them, as fossil fuel export revenues decline. Thus, the past remains effectively locked in.[15] Ironically, climate change itself will stand in the way of undoing the Soviet past and rationalizing the economy.

Not everything can be blamed on the Soviet planners. Geography itself has given Russia both a blessing and a curse, in the form of a treasure trove of natural resources in a climate that is poorly suited to anything else but their development and extraction. Russia's comparative advantage is inextricably tied to natural resources, and first and foremost to fossil fuels. As Putin himself observes, defending Russia's hydrocarbon model at a meeting with other world leaders, "Hydrocarbons are our comparative advantage!"[16] Yet the question is what Russia has done with the proceeds.

Finally, Russia's demography aggravates the situation. Its declining population—itself a legacy of the past—means that Russia does not have large reserves of cheap labor, as it did a century ago when Stalin forced millions of peasants off the farms to support his forced-draft industrialization. Russia's total fertility rate (the number of children the average woman has over a lifetime) remains too low to replenish the population, and the decline is only partially offset by migration, chiefly from Central Asia. The impact is especially severe on Russia's agriculture, as young people are leaving the countryside for the cities, a potential threat to an otherwise remarkable success story. A similar migration is taking place in the Far North, as Indigenous Peoples move to the handful of large cities such Yakutsk. The result is a growing demographic and ethnic imbalance, which imposes costs.

Yet arguably the greatest weight of the past is on Russia's culture and institutions. This is particularly evident in Russia's relations with the outside world. Climate change will make this problem even more difficult, as the next section will argue.

Russia and the Outside World

Climate change, combined with Russia's diminishing export income, will have several far-reaching consequences for Russia's position in the outside world. The first is that Russia will become relatively more isolated from the world economy. In contrast with China, it will continue to be less integrated. Finally, it is unlikely to play a leading role in the global diplomacy of climate change.

The Russian economy today is relatively open to the world compared to Soviet times, when the state practiced a systematic policy of economic isolation. In 2017, Russia's exports-to-GDP ratio was 26 percent, putting Russia at 115th place out of the 164 countries tracked by the World Bank.[17] However, since the beginning of the century the ratio of exports to GDP has steadily declined, from the high 30–40 percent range to the mid-20s. The Russian economy is evolving toward a greater share of consumer goods and services, which are mostly traded at home. Thus the Russian economy is less open to the world economy than it appears, and if exports of raw materials were to diminish without being offset by other exports, it would become apparent that in reality Russia's involvement with the world economy is declining, not growing.

Russia's recent politics push it in the same direction. Compare the evolution of Russia with that of China. Over the last thirty years, China has been transformed from a poor peasant economy, closed to the outside world, to a powerhouse of the global economy and a technological rival to the United States. During the same period, Russia has remained suspicious of integration, resistant to openness, ambivalent toward foreign investment, and isolated from major scientific and technological currents.[18] These tendencies have been reinforced under

Putin, and the state-capitalist system that he will bequeath to Russia is likely to remain as dependent on raw-materials exports as before. As Georgetown professor Harley Balzer has argued, China's integration with the world economy is *thick*; Russia's integration with the world economy is *thin*.[19]

Russia's integration with the world economy is also *negative,* in the sense that capital is leaving the country instead of coming in. Russia has been a net exporter of capital for virtually its entire post-Soviet history, as Russian private investors move their assets into safe havens abroad. (The one exception was the period from 2005 to 2007, on the eve of the global financial recession.) From 2015 to 2018, in the wake of US sanctions and low oil prices, outflows of direct investment hovered around 2 percent of GDP, while inflows were under 2 percent in most years.[20] Russia suffers in particular from the tendency of Russian companies and wealthy individuals to move their capital out of Russia. According to a recent study, "the wealth held offshore by rich Russians is about three times larger than official net foreign reserves, and is comparable in magnitude to total household financial assets held in Russia."[21] This points to a fundamental weakness—a lack of faith in Russia's future by those who would be in the best position to change it. Once again, the consequence is that Russia's ability to diversify its economy and adapt to climate change will be greatly diminished.

Russia's growing isolation is aggravated by the sanctions imposed by the United States and the European Union in the wake of the 2014 seizure of Crimea and Russia's role in eastern Ukraine. These are a further drag on the Russian economy, mainly through their negative impact on foreign investment. Unfortunately for Russia, the sanctions are likely to remain in place for the foreseeable future; indeed, they may worsen. In the United States, the Russia sanctions are now governed by acts of law instead of executive orders; in effect, Congress is now in control of sanctions policy. As shown by the famous example of the Jackson-Vanik amendment—an act of law that remained in place for over a generation—measures enacted by Congress against Russia are almost impossible to undo. The Russians are gloomily aware of this.

As its financial means diminish, Russia will have little incentive or capacity to play a leading role in the global diplomacy of climate change. It will continue to invoke its progress under the benchmark of its performance since 1990, and it will stress the role of its forests as absorbers of CO_2. But its overall contribution to cutting CO_2 emissions will be hardly more than symbolic, because the economics of energy inside Russia weigh heavily against any rapid transition, and Russia will have few resources to invest in the modernization of its inefficient industries. Russia will likely join the wave of formal decarbonization commitments being made by other governments, and it will multiply official climate strategies and doctrines promising long-range cuts in emissions, but to little actual effect. Russian companies, especially those with overseas assets, will advertise their green achievements, but much of that will amount to greenwash, aimed at preventing the delisting of Russian companies abroad and protecting their foreign assets. Russia will bargain hard to limit carbon taxes imposed on its exports, but its capacity to decarbonize its fossil fuel exports will remain limited. The one possible exception could be the development of hydrogen, if technological breakthroughs yet to come make the production and transportation of hydrogen feasible, in combination with natural gas. But a true "hydrogen revolution," which is at least two decades away in any case, is more likely to harm Russia's remaining energy exports than otherwise.

Russia: A Declining Great Power?

The era of globalization happened to coincide with the end of the Soviet Union and Russia's sudden reemergence into the world economy. Russia entered the twenty-first century much diminished compared its standing in Soviet times, when it boasted the world's second-largest economy and rivaled the United States in military technology and power. At the end of 1991, the Soviet Union abruptly lost half of its population, and Russia's borders receded to the historic core of the seventeenth century. Its economy plummeted; its living standards collapsed; and its educational and scientific systems dissolved. Despite an impressive recovery between 2000 and 2015, largely due to a long surge in oil prices and a prudent fiscal and budgetary policy, Russia's economic growth has stagnated since then and, I have argued in this book, will continue to do so.

Will Russia still be a great power by 2050? Professor Timothy Colton, one of the West's leading experts on Russia, put the question squarely in a recent elegant synthesis of Russia's situation.[22] In answer to his own question, Colton lists five features that would still qualify Russia as a great power. Significantly, four out of five have to do with military power—conventional weapons, nuclear weapons, space communications, and capacity for cyberwarfare. The fifth, Russia's long borders, could be arguably regarded as being as much as a liability as a source of strength. Nevertheless, "geopolitically," Colton concludes, "Russia is a bona fide great power."

This is undoubtedly correct as far as it goes. Yet much of Russia's military and technological capability is still based on assets inherited from the Soviet era. Despite the reforms undertaken since 2008 to modernize the military sector, it has not yet succeeded in developing a world-class new generation of weapons and space vehicles, except mainly as prototypes for display in parades.[23] Russia's crucial underlying weakness is its failure to keep up with the pace of scientific advance and technological innovation among the leading powers—much of which stems from civilian products, developed mainly in East Asia and exported throughout the world.[24] Military power ultimately depends on economic and technological strength. In these respects, Russia is already a declining power.[25]

Russia's Options at Mid-century

Readers may object that this picture of Russia in 2050 is too bleak by half, and I agree. It defies belief that a nation-state of 144 million talented, highly educated, and proud people will allow themselves to sink passively to second-rank status. In many respects Russia is a developed country (not a "developing" country, despite what many Western pundits and politicians routinely say) with an abundance of leading scientists, technologists, and entrepreneurs. Russia has been reborn many times in the past. It will be reborn again in the future. But on what basis?

The key is indeed high technology, as Putin never tires of saying. But for Russia to become a high-technology powerhouse, it must somehow "leapfrog" over its present liabilities while taking advantage of its remaining natural-resource income while it lasts. Leapfrogging will require three things: a reopening to the world economy in the form of "thicker" integration, the creation of an investment climate that will halt the exodus of capital and attract it back to Russia, and above all, the offer of an appealing challenge to Russia's young generation—both at home and abroad—that will unleash its talent and enthusiasm.

By 2050 a new generation will have come to power in Russia. It will be the first generation to have grown up entirely in the post-Soviet era. In many ways it will be very different from any previous generation in Russia. On present trends, it will be more broadly educated, compared to the predominantly engineering-oriented skill base of the Soviet era. It will be vastly more international, both directly through personal travel and indirectly through the internet. It will take many personal freedoms for granted. And it will have grown up in a market economy, in which many of the essential institutional foundations, unlike the 1990s, have by now been laid.

Russians today understand that their future lies with the deployment of this human capital. This is hardly a new observation. "People are our second oil" has been a mantra among Russian leaders throughout the Putin period. But the phrase itself is revealing: the point is that people should be Russia's *first* oil, whereas under Putin they have been far down the list. The greatest need is to empower this new generation, by removing the obstacles that stand in the way of entrepreneurship and innovation—and above all loosening the grip of the all-powerful and corrupt bureaucracy and the security apparatus. It is a familiar list; reformers never tire of speaking of it. But it requires overcoming the ingrained tendency of Russian officialdom to make the state great by making the people small—and themselves wealthy.[26]

Climate change is not the cause of these problems; it is a catalyst. Through the internal and external costs it will increasingly impose on Russia, it will precipitate the end of Russia's hydrocarbon model, while denying it the revenues and resources it will need to bring about change. But climate change is also an opportunity. If it brings any benefit to Russia, it will be to galvanize Russian society into action. Russia's new generation will hope for a fresh start. But by 2050, it will face a great reckoning.

NOTES

ACKNOWLEDGMENTS

INDEX

NOTES

Introduction

1. For a candid and insightful analysis by James Henderson of the Oxford Institute of Energy Studies and Tat'iana Mitrova, head of the Energy Center at the Skolkovo Business School in Moscow, see their article "The Implications of the Global Energy Transition on Russia," in M. Hafner and S. Tagliapietra, eds., *The Geopolitics of the Global Energy Transition*, Lecture Notes in Energy 73 (Cham, Switzerland: Springer, 2020), 93–114.

2. Several coastal cities in Russia, notably Murmansk and Vladivostok, are built on high ground. But Russian climatologists expect that Saint Petersburg will experience serious flooding by the end of the century. See the overview of the impact of climate change on the Russian economy by Aleksey Kokorin, head of the World Wildlife Fund Russia, in Aleksandra Koshkina, "Teplovoi udar po ekonomike," *Profil'*, no. 16 (April 29, 2019), pp. 20–24.

3. See the discussion of Miami and New Orleans in Orrin H. Pilkey, Linda Pilkey-Jarvis, and Keith C. Pilkey, *Retreat from a Rising Sea: Hard Choices in an Age of Climate Change* (New York: Columbia University Press, 2016), chaps. 3 and 4.

4. Michel Wara of the California Wildfire Commission, interview with Chris Nelder, *The Energy Transition Show*, podcast no. 102, August 21, 2019, https://xenetwork.org/ets/episodes/episode-102-transition-as-wildfire-adaptation-in-california/.

5. I am indebted for these lines of thought to Kingsmill Bond and his colleagues in *The Speed of the Energy Transition*, a report published jointly in September 2019 by Carbon Tracker, Bloomberg New Energy Finance, and the Rocky Mountain Institute, https://www.carbontracker.org/reports/speed-of-the-energy-transition/. This view may seem overly optimistic to some readers, and should not underestimate the capacity of incumbents to resist financial signals through

interest-group politics, yet the lesson of the last two decades is that the former do ultimately prevail. But for a well-written contrary view with sobering implications, see Leah Cardamore Stokes, *Short-Circuiting Policy: Interest Groups and the Battle over Clean Energy and Climate Policy in the United States* (Oxford: Oxford University Press, 2020). This book recounts the multiple ways in which incumbents can delay the adoption of clean energy technologies and reverse advances already made. The advent of the Biden administration and the adoption of a committed climate change policy at the federal level may help to overcome these obstacles.

6. Both the IMF and the IEA have issued reports on the possible long-term effects of the virus for climate change. There will, of course, be many more to come.

7. The best evidence is the successive reports of the Intergovernmental Panel on Climate Change (IPCC). The latest IPCC report, called *AR6 Climate Change 2021,* is scheduled for release in October 2021, to be followed by its *6th Assessment Report,* scheduled for 2022; see https://www.ipcc.ch/assessment-report/ar6/. At the same time, I am mindful of the objections of many atmospheric physicists that rising CO_2 concentrations alone cannot account for the growing symptoms of climate change, especially if one considers that water and thin clouds are by far the most important agents of the greenhouse effect, whereas CO_2 concentrations are very small by comparison. There is clearly much that remains to be understood. Nevertheless, in this book I have taken the IPCC as my most reliable guide to the consensus view of the community of climate scientists.

8. Vaclav Smils argues this point persuasively in his *Energy and Civilization: A History* (Cambridge MA: MIT Press, 2017), pp. 388–397.

9. On the relative speeds of previous energy transitions, see the pathbreaking study by Reda Cherif, Fuad Hasanov, and Aditya Pande, "Riding the Energy Transition: Oil beyond 2040," IMF Working Paper WP/17/120, May 2017, https://www.imf.org/en/Publications/WP/Issues/2017/05/22/Riding-the-Energy-Transition-Oil-Beyond-2040-44932.

10. On this point, see the handbook by MIT economist Bruce Usher, *Renewable Energy: A Primer for the Twenty-First Century* (New York: Columbia University Press, 2019). Two especially strong points are his focus on the total costs of renewables (LCOE, or levelized cost of electricity) and his discussion of the possible future synergies between renewable power and electric vehicles.

11. This is one of the chief conclusions of BP's 2019 and 2020 outlooks. As BP's chief economist, Spencer Dale, sums them up, the main thing to watch is not electric vehicles, but electric power. See Dale's 2019 and 2020 presentations on the website of the Center for Strategic and International Studies, at https://www.csis.org/events/2019-bp-energy-outlook and https://www.youtube.com/watch?v=QhuI1_UwxWs.

12. *BP Energy Outlook: 2019 Edition,* p. 23, https://www.bp.com/content/dam/bp/business-sites/en/global/corporate/pdfs/energy-economics/energy-outlook/bp-energy-outlook-2019.pdf.

13. Ipek Gençsü et al., *G20 Coal Subsidies: Tracking Government Support to a Fading Industry* (London: Overseas Development Institute, June 2019), https://www.odi.org/sites/odi.org.uk/files/resource-documents/12744.pdf. The report emphasizes that this number significantly underestimates the true total.

14. See the interview with Lauri Myllyvirta, one of the West's leading experts on the politics of Chinese coal, on the *Energy Transition Show,* episode #138, January 26, 2021, https://xenetwork.org/ets/episodes/episode-138-transition-in-china/. A summary of the main points by the same author is Lauri Myllyvirta, "Analysis: Will China Build Hundreds of New Coal Plants in the 2020s?," Carbon Brief, March 24, 2020, https://www.carbonbrief.org/analysis-will-china-build-hundreds-of-new-coal-plants-in-the-2020s.

15. Based on the EIA's "reference" case, a "business as usual" trend estimate.

1. The Politics of Climate Change in Russia

1. For recent examples, drawn from the month of December 2020 alone, see Aleksandr Kislov of the Moscow University Geography Department and Vladimir Semenov of the Institute for Atmospheric Physics (https://www.rbc.ru/rbcfreenews/5e672c469a7947421e50447c?from=materials_on_subject) and Andrei Kiselev of the Voyeikov Geophysical Observatory (https://rg.ru/2020/01/12/reg-urfo/uchenye-obiasnili-pochemu-klimat-rossii-tepleet-bystree-vseh-v-mire.html).

2. See, for example, an article coauthored by Oleg Anisimov, who is in many respects the dean of Russian Arctic climate studies, which argues against the concept of a "methane bomb" on the basis of new research and a systematic reanalysis of available data. Oleg Anisimov and Sergei Zimov, "Thawing Permafrost and Methane Emission in Siberia," *Ambio* (an online publication of the Royal Swedish Academy of Sciences), November 2020, https://doi.org/10.1007/s13280-020-01392-y. There is indeed a growing phenomenon of "methane craters" in the permafrost, but those are smaller and more localized events.

3. Speech by Rosneft CEO Igor Sechin at the XIII Eurasian Economic Forum in Verona, Italy, reported in Vladimir Polkanov, "Igor' Sechin prizval poiti po kitaiskomu puti," *Nezavisimaia gazeta,* October 23, 2020, https://www.ng.ru/economics/2020-10-22/8_7997_164322102020.html.

4. A recent example is the three-day seminar conducted in December 2020 by Iury Mel'nikov of the Skolkovo energy program, featuring Russian experts on topics ranging from permafrost to hydrogen, in dialogue with European counterparts (https://www.skolkovo.ru/programmes/eu-russia-climate-conference

/broadcast/). Skolkovo organizes regular "summer schools" for Russian audiences on various aspects of energy transition and their implications for Russia. For the 2019 edition, see https://www.youtube.com/watch?v=YzuwpqHNc0o&list=PLZ WnvmqHFxphEDUmGXmcjHp2ujmO3Hzml.

5. See, for example, the extensive joint report of the Skolkovo Energy Center and the Institute of Energy Studies of the Russian Academy of Sciences, "Global and Russian Energy Outlook 2019," available on the website of the Skolkovo Energy Center, https://www.eriras.ru/files/forecast_2019_en.pdf.

6. Maksim Reshetnikov, born in 1979, was formerly the governor of Perm' Kray (2017–2020). Prior to that he served in the Moscow government under Mayor Sergei Sobianin (2009–2017). In January 2020 he was named minister of economics as part of the incoming government of Prime Minister Mikhail Mishustin. Under Reshetnikov the Ministry of Economic Development has been the primary campaigner for tighter emissions controls, against the opposition of the Ministry of Industry and Trade and the Ministry of Energy.

7. Aleksandr Kozlov, born in 1981, was previously governor of Amur Oblast and comes from an early career in the local coal industry. However, from 2018 to 2020 he was minister of development of the Far East. During that time the remit of the ministry was broadened to include the Arctic, at a time of rising priority of the Arctic in Russian policy. He was replaced at that post by Aleksey Chekunkov.

8. Chekunkov, born in 1980 and a graduate of the prestigious Moscow State Institute of International Relations (MGIMO), pioneered venture capital as a private entrepreneur before joining VEB.RF as the general director of the Fund for the Development of the Far East and the Arctic, where he served between 2014 and 2020.

9. Sorokin, a former vice president of Morgan Stanley and a Western-educated expert on international finance, was the principal author of the controversial "Energy Strategy," discussed in Chapter 2. For a brief biography of Sorokin, see https://minenergo.gov.ru/node/10798. Krutikov became for a time the public face of Russia's expanded Arctic strategy, a policy that is likely to continue under the new minister, Aleksey Chekunkov. For *Kommersant's* coverage, see https://www.kommersant.ru/doc/4783410.

10. See the discussion of Novak's promotion in *Neftegazovaia Vertikal'*, no. 1 (2021), http://www.ngv.ru/magazines/article/povyshenie-aleksandra-novaka/. For a brief biographical sketch, see http://government.ru/en/gov/persons/187 /events/.

11. On "peak oil demand" and its possible consequences for Russia, Novak appears to be at one with the views and concerns of the Ministry of Finance. In December 2020, a deputy minister of finance, Vladimir Kolychev, stated in an interview with Bloomberg, "The peak of consumption may have already passed." See Anna Andrianova and Evgenia Pismennaya, "Russia Starts Preparing for

Life after Peak Fossil Fuels," *Bloomberg Green,* December 5, 2020, https://www
.bloomberg.com/news/articles/2020-12-05/russia-starts-preparing-for-life-after
-peak-fossil-fuels. There is a running disagreement on this point between the
Ministry of Finance and the Ministry of Economic Development, on the one hand,
and the Ministry of Energy on the other. Yet, significantly, the Ministry of Energy
too accepts the peak oil demand narrative but believes that the peak will not
come before 2045.

12. Novak is an interesting, indeed a unique figure. A Ukrainian by birth,
Novak moved with his family as a child to the "nickel capital" of Norilsk on the
Arctic coast of Siberia (a site that, as we shall see in later chapters, has become a
household word in Russia as the location of the worst Russian environmental disaster of recent times and a symbol of the growing impact of climate change).
There Novak began his career as an ordinary worker, rising from the shop floor
of the Norilsk metallurgical combine while earning degrees in metallurgy and
economics at the local Financial Institute. By the age of twenty-nine he was vice
president of the company; in his early thirties he became deputy mayor of Norilsk, then vice-governor of Krasnoyarsk Province. From there, at age thirty-seven,
Novak was named deputy minister of finance in Moscow, and four years later he
became minister of energy. For a short biography, see Aleksandr Borisov, "U piati
ministerstv Rossii poiavilis' novye rukovoditeli," *Sankt-Peterburgskie vedomosti,*
November 20, 2020, https://sanktpeterburg.bezformata.com/listnews/rossii
-poyavilis-novie-rukovoditeli/88989259/. (Among other interesting details, Novak,
who is 6 feet, 3 inches tall, played on his high school basketball team in Norilsk.)
On his recent interventions in defense of renewables, see Polina Smertina and
Tat'iana Diatel, "Zelenaia energetika ne rezhetsia na glaz," *Kommersant,* January 20,
2021, https://www.kommersant.ru/doc/4653626.

13. Novak appears to divide the "environmental account" at the top of the
government with the newly promoted Viktorya Abramchenko, with Novak
taking responsibility for climate change while Abramchenko oversees "classic"
environmental-quality issues such as waste disposal. See the interview with
Abramchenko in *Kommersant,* January 12, 2021, https://www.kommersant.ru
/doc/4640076. Abramchenko is very much the junior partner. Her background and
expertise lie in creating cadastres and registering land use, especially agricultural—
in other words, counting things. One Russian response to the threat of the
European Union's carbon export tax has been has been to recount the trees
and increase their claimed absorptive capacity. Abramchenko is responsible
for this effort.

14. Ruslan Edel'geriev reportedly comes from the same village as Kadyrov and
is said to be the son of Kadyrov's childhood teacher. See Dar'ia Zelenskaia,
"'Bratu' Kadyrova doverili rossiiskii klimat," *Moskovskii komsomolets,* June 26,

2018, p. 1. For Edel'geriev's official Kremlin biography, see http://www.kremlin.ru/catalog/persons/571/biography.

15. See the report on Edel'geriev's private meeting in https://meduza.io/en/news/2020/02/06/kremlin-climate-representative-acknowledges-heating-effects-in-press-conference-following-meeting-with-activists.

16. Angelina Davydova, "Debaty o klimate stanoviatsia zharche," *Kommersant,* July 31, 2020, accessed via East View at https://dlib-eastview-com.proxy.library.georgetown.edu/browse/doc/60882277. See the regular reports of the Interagency Working Group on the Kremlin website—for example, at http://kremlin.ru/events/administration/64677. Consequently, he was the logical protocol choice to represent the government at the meeting with John Kerry.

17. VEB.RF is the descendant of the Soviet-era Foreign Economic Bank, which was renamed VEB in the 1990s. In early 2021 VEB was renamed VEB.RF and has been reconfigured as Russia's main high-tech venture capital provider. In the process it has absorbed Rusnano, as well as the Skolkovo Foundation. At this writing it is unclear how this will affect the independence of the Skolkovo Energy Center.

18. Gleb Maikov, "Chubais predstavliaet," *Nasha versiia,* no. 48 (December 20, 2020), https://versia.ru/na-novoj-dolzhnosti-privatizator-chubajs-potesnit-sergeya-ivanova-i-ruslana-yedelgerieva. Chubais's full title reads, "Special Representative of the President of the Russian Federation for Ties with International Organizations for the Attainment of the Goals of Sustainable Development." There are only two "special representatives" in the Presidential Administration. The other is Sergei Ivanov, who has held the position since 2016. At the time, Ivanov's appointment was considered a demotion; he had previously been the head of the Presidential Administration.

19. The meeting was reported in "Klimaticheskaia diplomatiia vozvrashchaetsia," *Kommersant,* February 9, 2021, https://www.kommersant.ru/doc/4721165?from=main_8. It is clear that Edel'geriev has come sharply up the learning curve, and he has gained self-assurance along the way. See for example, his extended interview with Angelina Davydova, *Kommersant*'s noted expert on climate issues, "Biznes dekarboniziruetsia na bumage i v korporativnykh otchetov," *Kommersant,* February 15, 2021, https://www.kommersant.ru/doc/4691458.

20. "Poslanie Prezidenta Federal'nomu Sobraniiu," April 21, 2021, http://kremlin.ru/events/president/news/65418.

21. For an overview of the early development of Western thinking on global warming, see Daniel Yergin, *The Quest: Energy, Security, and the Remaking of the Modern World* (New York: Penguin Press, 2011), chap. 22. On Keeling's Curve, see pp. 442–443.

22. An excellent review of the early history of discourse on climate change in the Soviet Union, as well as Budyko's own role, is Katja Doose and Jonathan Old-

field, "Natural and Anthropogenic Climate Change Understanding in the Soviet Union, 1960s–1980s," in Marianna Poberezhskaya and Teresa Ashe, eds., *Climate Change Discourse in Russia: Past and Present* (New York: Routledge, 2019), pp. 17–31.

23. Doose and Oldfield, "Natural and Anthropogenic Climate Change Understanding," pp. 21–24. See also Oleg Anisimov and Vasiliy Kokorev, "Cities of the Russian North in the Context of Climate Change," in Robert W. Orttung, ed., *Sustaining Russia's Arctic Cities: Resource Politics, Migration, and Climate Change* (New York: Berghahn Books, 2017), p. 143.

24. One prominent example is Oleg Anisimov, a former student of Budyko and an early participant in the IPCC, who has made a specialty of studying the impact of climate change in the Arctic. See, for example, Oleg Anisimov and Robert Orttung, "Climate Change in Northern Russia through the Prism of Public Perception," *Ambio* 48 (2019): 661–671, https://doi.org/10.1007/s13280-018-1096-x.

25. The Ministry of Economic Development has changed its name several times as its various functions have been reshuffled. Between 2000 and 2008 it was called the Ministry of Economics and Trade. In 2008 it transferred its trade responsibilities to a new Ministry for Trade and Industry and was given its present name, Ministry of Economic Development. For the sake of clarity, I have used its present name throughout the book.

26. See, for example, Ivan Kostromin, "Klimat: Kioto razdelil Putina i ego sovetnika," *Novaia gazeta*, May 27, 2004, p. 19. (This source is no longer available on the website of the newspaper, but can be obtained from the author via East View at Georgetown University.) An extreme libertarian, Illarionov favored setting up a carbon market as the best means of curtailing emissions, outside the framework of Kyoto's mandated emissions targets (see also Aleksey Shapovalov, "Globalizatsiia: Kiotskii protokol meshaet udvoeniiu VVP," *Kommersant*, October 4, 2003, p. 5, https://www.kommersant.ru/doc/416871). In this respect Illarionov was far ahead of the Russian consensus. Interestingly, however, his proposal for a carbon market was picked up by Medvedev's climate advisor Aleksandr Bedritsky, but it continues to be strongly opposed by most of Russian industry.

27. This at least was the widespread speculation in the West at the time. See, for example, Peter Lavelle, "EU-Russia Trade Horses," *UPI Defense News*, May 21, 2004, https://www.upi.com/Defense-News/2004/05/21/Analysis-EU-Russia-trade-horses/86361085167739/.

28. Kremlin website, 2005, various locations. Throughout 2005 and 2006, Putin held frequent meetings, either in person or by telephone, with both Chirac and Schroeder (see, for example, http://kremlin.ru/events/president/news/32976). Unfortunately, these meetings, while still listed on the Kremlin website, are no longer publicly available.

29. http://kremlin.ru/events/president/transcripts/23368. (The same is true of this transcript as well.)

30. The Russian Constitution of 1996 provides that the president can serve only two terms. Accordingly, at the end of 2008, Putin stepped down in favor of his protégé Dmitry Medvedev and served four years as prime minister. But in 2012 Putin ran again for president, on the grounds that the constitution only banned two *consecutive* terms. On that reading, Putin, who then won a fourth term in 2018, would remain president. Finally, in late 2020, Putin arranged for amendments to the constitution, which will enable him to remain in office indefinitely.

31. Medvedev created a Commission on Modernization and Technological Development of Russia, which met frequently to discuss a wide range of topics. In June 2011 the topic was ecology (see Medvedev's remarks at http://kremlin.ru /events/president/news/11755). After Putin's return to the presidency the commission soon lapsed.

32. By 2017, energy intensity of GDP was only 10 percent lower than in 2007. I. Bashmakov, "What Happens to the Energy Intensity of Russia's GDP?," *Ecological Bulletin of Russia* 7, no. 8, cited in James Henderson and Tatiana Mitrova, "Implications of the Global Energy Transition on Russia," in M. Hafner and S. Tagliapietra, eds., *The Geopolitics of the Global Energy Transition,* p. 106, Lecture Notes in Energy 73 (Cham, Switzerland: Springer, 2020). In 2018 the target was scaled down to 9.4 percent and funding was discontinued. Ministry of Economic Development of the Russian Federation, "State Report on the State of Energy Savings and Energy Efficiency in the Russian Federation in 2017" (in Russian) (Moscow, 2018), cited in Henderson and Mitrova, "Implications of the Global Energy Transition on Russia," p. 113. In 2021, in a lengthy and candid interview (see note 16 above), Edel'geriev acknowledged that the 2007 decree had been a failure.

33. In the early 2000s, for example, the director of Rosgidromet, Roman Vil'fand, stated, "Humankind in this process has a modest role. . . . It is not within our powers to stop the changes." Elana Wilson Rowe, *Russian Climate Politics: When Science Meets Policy* (New York: Palgrave Macmillan, 2013), pp. 36, 45.

34. *Otsenochnyi doklad ob izmeneniiakh klimata i ikh posledstviiakh na territorii Rossiiskoi Federatsii,* 2 vols. (Moscow: Rosgidromet, 2008), vol. 1, part VI. Both volumes, together with an English-language summary, are available on the website of Rosgidromet, http://climate2008.igce.ru/v2008/htm/index00.htm. In 2014 Rosgidromet published a second massive overview, summarizing the efforts under way at that time (*Vtoroi ostenochnyi doklad Rosgidrometa ob izmeneniiakh klimata i ikh posledstviiakh na territorii Rossiiskoi Federatsii,* http://downloads.igce .ru/publications/OD_2_2014/v2014/htm/). Since 2008 Rosgidromet has been consistently aligned with the views of the IPCC. See the interview with the then-director of the Rosgidromettsentr, Roman Vil'fand, in Iuliia Tutina and Roman

Vil'fand, "Chto tvoritsia s pogodoi?," *Argumenty i Fakty,* no. 19 (2017). By this time Vil'fand's public views had evolved considerably, and he had become a strong voice warning of the effects of man-made climate change.

35. The full text of the Climate Doctrine is available on the website of the Russian president, http://www.kremlin.ru/events/president/news/6365.

36. Irina Granik, "U Dimitriia Medvedeva poiavilsia sovetnik po pogode," *Kommersant,* November 28, 2009, p. 3.

37. Marianna Poberezhskaya, *Communicating Climate Change in Russia: State and Propaganda* (New York: Routledge, 2016), pp. 8off.

38. Ellie Martus, "Russian Industry Discourses on Climate Change," in Poberezhskaya and Ashe, *Climate Change Discourse,* pp. 97–112. In contrast, Martus notes that the metals and mining industry has played a more active part in policy debates on climate change, both for and against the government's policies. Rusal has been a prominent supporter of international initiatives to curb emissions, including support for a price on carbon. See also Vasilii Stolbunov, "Biznes osvaivaet bezuglerodnye tekhnologii," *Nezavisimaia gazeta,* December 18, 2018, p. 4, https://www.ng.ru/economics/2018-12-17/4_7464_19451712.html.

39. The shortage of "legacy gas" in the first half of the 2000s is discussed in Thane Gustafson, *The Bridge: Natural Gas in a Redivided Europe* (Cambridge MA: Harvard University Press, 2020). By 2006 it led to the decision to begin developing the next generation of gas supply from the Yamal Peninsula. But in the early 2000s, as Putin came into office, the perceived shortage was acute, although the amount of associated gas lost through flaring may have been exaggerated.

40. This point is discussed in Chapter 3.

41. The website of the Global Gas Flaring Reduction Partnership (GFRP) can be accessed at https://www.worldbank.org/en/programs/gasflaringreduction. The policy has been effective up to a point: in 2017 the oil industry recovered more than 85 billion cubic meters of associated gas, but continued to flare about 13 billion cubic meters, still leaving it short of the 95 percent limit. In 2020 the oil companies came under increased government pressure, as fines were raised again. In February 2021, in a meeting with Putin at the Kremlin, Rosneft chairman Igor Sechin reported that Rosneft had achieved 83 percent recovery of associated gas and would shortly reach the mandated level of 95 percent. Of all the various new "green" initiatives by Rosneft that Sechin reported to the president (including a plan for a wind farm at Vostok Oil), the one aspect that Putin showed interest in was the recovery of associated gas. This is presumably because it is the most observable part of Rosneft's carbon footprint, and as such it would be a prime target of the EU's proposed carbon export tax. For a partial transcript of the meeting, see http://kremlin.ru/events/president/news/65000. For the World Bank's latest Gas Tracker report, April 2021,

see https://thedocs.worldbank.org/en/doc/1f7221545bf1b7c89b850dd85cb409b0
-0400072021/original/WB-GGFR-Report-Design-05a.pdf. Since 2012, the GFRP
has been tracking gas flaring by satellite.

42. Or in technical terms, the "levelized costs" of solar and wind power had
fallen below "grid parity." On the sharp decline in levelized costs for wind and solar
in the 2010s and its implications, see, for example, a convenient overview by Bruce
Usher, *Renewable Energy: A Primer for the Twenty-First Century* (New York:
Columbia University Press, 2019).

43. The text of Putin's speech appears on the president's website at http://
kremlin.ru/events/president/news/60707.

44. The duties and composition of the Working Group can be found on the
Kremlin website at http://kremlin.ru/structure/administration/groups#institution
-1003.

45. The text of the Strategy can be found on the Kremlin website at http://
kremlin.ru/acts/bank/41879/page/2. The Strategy is a very useful document and
includes a wealth of statistics on the state of the Russian environment. But it is
largely confined to traditional environmental problems such as waste and pollution, on which it paints a remarkably negative picture. It has only a few passing
references to climate change.

46. The full document can be accessed at http://static.government.ru/media
/files/OTrFMr1Z1sORh5NIx4gLUsdgGHyWIAqy.pdf. For a helpful commentary,
see Pavel Devyatkin, "Russia Unveils Climate Change Adaptation Plan," *High North
News,* January 8, 2020, https://www.highnorthnews.com/en/russia-unveils-climate
-change-adaptation-plan.

47. The most significant of the three Arctic documents was the National Action
Plan on Measures to Adapt the Economy and the Population to Climate Change.
For the full text, see http://static.government.ru/media/files/OTrFMr1Z1sORh5N
Ix4gLUsdgGHyWIAqy.pdf. In the preparation of these documents the Ministry
of Economic Development has played the lead role. For a helpful guide, see Elizabeth Buchanan, "The Overhaul of Russian Strategic Planning for the Arctic
Zone to 2035," NATO Defense College, Russian Studies Series 3/20, May 19, 2020,
https://www.ndc.nato.int/research/research.php?icode=641; and Buchanan, "Russia's Updated Arctic Strategy," *High North News,* October 28, 2020, https://www
.highnorthnews.com/en/russias-updated-arctic-strategy-new-strategic-planning
-document-approved. Elizabeth Deakin is a lecturer at Deakin University, part of
the Australian War College. She has made a specialty of Russian Arctic policy.

48. See Putin's speech at the October 2019 Russian Energy Week, accessible on
the Kremlin website at http://kremlin.ru/events/president/news/61704. His
speech was especially significant because it was delivered in front of a panel that
included the CEOs of BP, ExxonMobil, and OMV.

49. The full transcript of the Q&A session appears on the presidential website, http://kremlin.ru/events/president/news/60795, June 20, 2019. In the "Direct Line" program just mentioned, solid waste was the only environmental question raised by his audience. There was no "Direct Line" meeting in 2020, but it was folded into Putin's annual press conference in December 2020.

50. The full text of the Q&A session appears on the presidential website at http://kremlin.ru/events/president/news/64671.

51. The full text of Putin's speech appears on the presidential website at http://kremlin.ru/events/president/news/60961.

52. Duma website, https://sozd.duma.gov.ru/bill/1116605-7.

53. Putin's speech to the Global Manufacturing and Industrialization Summit, July 10, 2019. The full transcript can be found on the presidential website at http://kremlin.ru/events/president/news/60961.

54. Full transcript at http://kremlin.ru/events/president/news/60961.

55. The transcript of Putin's speech is accessible on the presidential website at http://kremlin.ru/events/president/news/60250.

56. "Rossiia zovet," October 2020, http://kremlin.ru/events/president/news/64296.

57. See the Kremlin website, http://kremlin.ru/events/president/news/64261, October 2020. As noted earlier, this view is contested by some Russian climate scientists, notably Oleg Anisimov. See note 2 above.

58. For Ivanov's interview on the Russia Today television channel, see "Khotite, raskroiu gosudarstvennuiu tainu?," April 26, 2017, https://russian.rt.com/russia/article/382903-sergei-ivanov-intervju. Many thanks to Professor Vadim Grishin for bringing Ivanov's present role to my attention.

59. Angelina Davydova and Evgeniia Kriuchkova, "Klimat—eto ne strashno," *Kommersant Daily*, July 14, 2016, https://www.kommersant.ru/doc/3037549.

60. Oleg Nikiforov, "Teoriia vliianiia cheloveka na temperaturu Zemli kak vozmozhnaia statisticheskaia oshibka," *Nezavisimaia gazeta*, April 9, 2019, p. 12, https://www.ng.ru/energy/2019-04-08/12_7551_teoria.html.

61. Ananskikh continues to oppose the Ministry of Economic Development's plans to impose limits on greenhouse gas emissions from the energy industry. See the *Parliamentary Gazette* for October 3, 2019, at https://www.pnp.ru/social/biznes-mogut-zastavit-poschitat-vybrosy-parnikovykh-gazov.html.

62. Oleg Nikiforov, "Rossiia lish' prisoedinilas' k parizhskomu soglasheniiu, no ne ratifitsirovala," *Nezavisimaia gazeta*, October 8, 2019, pp. 9–10, https://www.ng.ru/energy/2019-10-07/9_7695_agreement.html.

63. Dmitrii Butrin and Aleksei Shapovalov, "Uglerodnye nalogi poshli na vybros," *Kommersant*, October 17, 2019, p. 1, https://www.kommersant.ru/doc/4127113.

64. Angelina Davydova and Aleksei Shapovalov, "Rossii propisali nizkou-glerodnoe budushchee," *Kommersant,* March 23, 2020, https://www.kommersant.ru/doc/4299377. See also Alina Fadeeva, "U Rossii poiavilsia plan po snizheniiu vybrosov parnikovykh gazov do 2050 goda," *RBC,* March 23, 2020, https://www.rbc.ru/business/23/03/2020/5e73c8739a7947f53f4f3a06. For another detailed description, see Angelina Davydova, "Bezuglerodnoe budushchee Rossi vy-gladit spornym," *Kommersant,* April 17, 2020, https://www.kommersant.ru/daily/125238.

65. https://tass.ru/ekonomika/9915107.

66. European Commission, "Financing the Recovery Plan for Europe," https://euagenda.eu/publications/financing-the-recovery-plan-for-europe. For background see Andrei Marci et al., "REPORT: Border Carbon Adjustments in the EU: Issues and Options," European Roundtable on Climate Change and Sustainable Transition (ECRST), September 30, 2020, https://ercst.org/border-carbon-adjustments-in-the-eu-issues-and-options/.

67. Ruslan Edel'geriev, "'Tsena na uglerod' kak instrument ekonomicheskoi i ekologicheskoi politiki," *Kommersant,* June 6, 2020, https://www.kommersant.ru/doc/4377361?from=doc_vrez.

68. Anna Fadeeva, "KPMG otsenila ushcherb dlia Rossii ot vvedeniia uglerodnogo naloga v ES," *RBC,* July 7, 2020, https://www.rbc.ru/business/07/07/2020/5f0339a39a79470b2fdb51be.

69. Angelina Davydova and Aleksei Shapovalov, "Rossii propisali nizkou-glerodnoe budushccee," *Kommersant,* March 23, 2020, https://www.kommersant.ru/doc/4299377.

70. This proposal comes from Deputy Prime Minister Viktorya Abram-chenko, whose primary expertise, as noted earlier, is in creating land cadastres. See her interview with Aleksei Shapovalov, "Ia za khoroshii sovetskii Gosplan," *Kommersant,* January 12, 2021, https://www.kommersant.ru/doc/4640076. A new law on forests embodying these ideas has been put before the Duma. Edel'geriev, while not opposing such proposals frontally, acknowledges European skepticism toward Russian promotion of its forests.

71. Angelina Davydova, "Na Sakhaline proeksperimentiruiut s uglerodom," *Kommersant,* March 2, 2021, https://www.kommersant.ru/doc/4711442.

72. Fadeeva, "KPMG otsenila ushcherb dlia Rossii."

73. "Poslanie Prezidenta Federal'nomu Sobraniiu," April 21, 2021.

74. For an on-the-ground opinion from Deutsche Welle's Moscow correspondent, see Andrey Gurkov, "Russia Frozen on Climate Change," *Deutsche Welle Online,* July 6, 2019, https://www.dw.com/en/opinion-russia-frozen-on-climate-change/a-49499528.

75. The questionnaire allowed for multiple answers and therefore the percentages do not sum to 100 percent. See https://www.levada.ru/2020/01/23/problemy-okruzha yushhej-sredy/. The latest Levada Center poll appeared in September 2020.

76. "Izmenenie klimata i kak s nim borot'sia," Levada Center, September 2020.

77. "Izmenenie klimata i kak s nim borot'sia," Levada Center, September 2020.

78. On this point see Igor Bashmakov, general director of the Center for Energy Efficiency 21st Century, in "Strategiia nizkouglerodnogo razvitiia rossiiskoi eko- nomiki," *Voprosy ekonomiki,* no. 7 (2020), p. 52, citing the 2019 Levada Center poll.

79. Alla Baranovsky-Dewey, "Why Russia's Garbage Protests Turned Violent," *Washington Post,* August 19, 2019, https://www.washingtonpost.com/politics/2019 /08/19/russias-garbage-protests-turned-violent-what-happened-lebanon-helps -explain-these-demonstrations/.

80. Dina Nepomnyaschyaya, "V pozharakh v Rossii sgorelo bolee 2.5 mln gek- tarov lesa," *Rossiiskaia Gazeta,* August 9, 2016, https://rg.ru/2019/08/09/reg -sibfo/v-pozharah-v-rossii-sgorelo-bolshe-25-mln-ga-lesa.html.

81. "Sibirskie pozhary do DiCaprio doveli: Socseti goriat, vlast' tushuetsia," *BBC News Russkaia Sluzhba,* July 31, 2019, https://www.bbc.com/russian/other -news-49181994.

82. Tat'iana Mitrova, "Energoperekhod i riski dlia Rossii," *Neftegazovaia vertikal',* no. 6, 2021, http://www.ngv.ru/magazines/article/energoperekhod-i-riski-dlya-rossii/.

2. The Twilight of Russian Oil?

1. I am indebted to Christopher Weafer, Keith King, John Webb, Vadim Grishin, and many other friends and colleagues, for their kind comments on earlier drafts of this chapter.

2. For a picturesque but not totally reliable account of Sechin's role and the rela- tionship between Putin and Sechin, see (with caution) the Sechin biography in the Lenta archive, at https://lenta.ru/lib/14160890/full.htm#108. There is reason to believe that Sechin successfully opposed Putin's plan to fold Rosneft into Gaz- prom. This episode is discussed in Thane Gustafson, *The Bridge: Natural Gas in a Redivided Europe* (Cambridge, MA: Harvard University Press, 2020), pp. 287–288.

3. There has been a lively debate over the years (in the West, at any rate) over whether Putin and Sechin actually wrote their own dissertations. Leaving that question aside, we might perhaps simply note that both Putin and Sechin report- edly attended a running roundtable on energy issues conducted by Vladimir Litvenenko of the Saint Petersburg Mining Institute, where they were presumably exposed to a wide range of energy issues. Litvenenko today remains part of the circle of favored Saint Petersburg personalities from both men's earlier days.

4. The fullest exposition of Sechin's views comes in a signed article, "Tsena nestabil'nosti," *Ekspert,* no. 45–46 (November 2019): 26–31. Sechin is especially critical of electric vehicles. The carbon footprint of electric vehicles, he writes, is 20 to 60 percent higher than that of conventional automobiles. A full transition to EVs will translate into a 30 percent increase in power demand worldwide, much of which will come from coal-fired power plants. Renewables are also losers: their advantages are overly idealized, Sechin argues, while those of traditional fuels are ignored.

5. The latest energy scenarios of BP and the similar projections of Equinor, from the fall of 2020, are available on the website of the Center for Strategic and International Studies (CSIS). For BP, see the presentation by Spencer Dale, chief economist of BP, https://www.csis.org/events/online-event-bp-energy-outlook -2020. For Equinor, see the presentation by Eirik Wærness, chief economist of Equinor, https://www.csis.org/events/online-event-equinor-energy-perspectives -2020.

6. See US Energy Information Administration, "OPEC+ Agreement to Reduce Production Contributes to Global Oil Market Rebalacing," September 23, 2020, https://www.eia.gov/todayinenergy/detail.php?id=45236.

7. Sechin's opposition was so public that some in the Russian media mockingly speculated whether Sechin had become, in effect, an "oil dissident." See Oleg Kashin, "Gosudarstvo i oligarkhiia: Prevratilsia Sechin v Khodorkovskogo?," *Republic (Slon),* March 20, 2020, https://republic.ru/posts/96205. Note in passing that the quotas as finally agreed upon apply only to production, even though, paradoxically, it is exports that directly influence the global oil market fundamentals and prices. So as long as Russia is within its production quota, it can export as much oil as it wants. And exports of refined products are not covered by the OPEC+ deal at all. (My thanks to Vitaly Yermakov of the Oxford Institute of Energy Studies and the Russian Higher School of Energy Economics for reminding me of these points. I am grateful to Dr. Yermakov for his critique of an earlier draft of this chapter.)

8. These are the projections from the reference scenarios of BP and Exxon-Mobil; they are at the high end of the range of estimates from the oil industry.

9. Spencer Dale and Bassam Fattouh, "Peak Oil Demand and Long-Run Oil Prices," Oxford Institute for Energy Studies (OIES), *Energy Insight,* no. 25, January 2018, https://www.oxfordenergy.org/publications/peak-oil-demand-long-run -oil-prices/. Spencer Dale is the chief economist of BP; Bassam Fattouh is director of OIES.

10. This is not necessarily good news for oil. Much of the demand for feed-stocks for the chemical industry consists of light fractions, especially so-called natural gas liquids (NGLs), a by-product of gas production. In this market the

United States is well positioned, thanks to its abundance of cheap "shale gas" and a well-established infrastructure for processing NGLs into plastics. Russia historically has paid little attention to NGLs, mainly exporting them to European plants instead of processing them itself.

11. On the comparative speed of energy transitions, see a thought-provoking paper by Reda Cherif et al., "Riding the Energy Transition: Oil beyond 2040," International Monetary Fund, Working Paper 17/120, May 22, 2017, https://www.imf.org/en/Publications/WP/Issues/2017/05/22/Riding-the-Energy-Transition-Oil-Beyond-2040-44932.

12. The fullest statements of the fast-transition narrative come from Climate Tracker and Bloomberg New Energy Finance (BNEF), a group of analysts who combine a focus on financial trends with energy economics. See particularly two reports: Kingsmill Bond et al., "The Speed of the Energy Transition: Gradual or Rapid Change?," White Paper prepared for the 2020 World Economic Forum, September 2019, http://www3.weforum.org/docs/WEF_the_speed_of_the_energy_transition.pdf; and Bond et al., "Decline and Fall: The Size and Vulnerability of the Fossil Fuel System," Carbon Tracker, June 2020, https://carbontracker.org/reports/decline-and-fall/.

13. The discussion that follows is based mainly on three sets of scenarios. The first comes from the International Energy Agency, contained in its annual *World Economic Outlooks.* The second is the various scenarios published by the major oil companies; I have relied chiefly on those from BP, ExxonMobil, Shell, and Equinor. The third is the fast-transition scenarios developed by various consultancies, such as BNEF and Carbon Tracker. All three are based on essentially the same methodology and attempt to take into account trends in population growth, technological progress, and political regulation. Each group of scenarios contains three broad "stories of the future," variously named, but all consisting essentially of a "business as usual" scenario, a "rapid transition" scenario, and a "reference" scenario, which might be freely translated as "our most plausible pathway."

14. See the presentation by Spencer Dale, the chief economist of BP, to CSIS, September 2020, https://www.csis.org/events/online-event-bp-energy-outlook-2020. BP's new scenarios are available at https://www.bp.com/en/global/corporate/energy-economics/energy-outlook.html. BP's corporate strategy, along with that of Equinor, has swung radically in the direction of renewables; indeed, the two companies have recently formed a partnership to develop renewables on a fast track. See https://www.bp.com/en/global/corporate/news-and-insights/reimagining-energy/bp-makes-first-move-into-offshore-wind.html. But Shell and Total, in their latest projections, also predict an earlier peak in oil demand.

15. Russian Customs Service, http://customs.ru/folder/513. The totals for oil combine crude, condensate, and refined products; those for gas include both pipeline gas and liquefied natural gas.

16. See the year-end report from the Ministry of Finance, https://minfin.gov.ru /ru/statistics/fedbud/execute/?id_65=80041-yezhegodnaya_informatsiya_ob _ispolnenii_federalnogo_byudzhetadannye_s_1_yanvarya_2006_g. This number refers to total federal revenues from oil and gas, which presumably includes domestic taxes as well as export revenues.

17. Ministry of Finance, year-end report. In 2016, oil and gas revenues in the federal budget were 4.8 trillion rubles.

18. For this reason, I have based this analysis on 2019 (the last "normal year") rather than 2020. But the 2020 numbers are available from https://customs.gov .ru/statistic. For the 2020 numbers for the federal budget, see https://minfin .gov.ru/ru/statistics/fedbud/oil/ and https://minfin.gov.ru/ru/statistics/fedbud /execute/. In 2020 oil exports accounted for $72.4 billion, refined product exports for $45.3 billion, pipeline natural gas exports for $25.2 billion, and LNG exports for $6.7 billion (21 percent, 13 percent, 7 percent, and 2 percent, for a 43 percent share of total Russian exports).

19. US Central Intelligence Agency, *Soviet Energy Atlas* (Washington, DC, 1985), https://www.cia.gov/library/readingroom/docs/DOC_0000292326.pdf.

20. For background on the Soviet era, see Thane Gustafson, *Crisis amid Plenty: The Politics of Soviet Energy under Brezhnev and Gorbachev* (Princeton, NJ: Princeton University Press, 1989).

21. In 2020, Rosneft's production of crude oil and condensate was 180 mt, plus a small additional amount from Rosneft's subsidiary Bashneft, at 12.9 mt, out of Russia's total of 512.8 mt.

22. For a description of the battles for the oil industry during the first two decades of the post-Soviet period, see Thane Gustafson, *Wheel of Fortune: The Battle for Oil and Power in Russia* (Cambridge, MA: Harvard University Press, 2012).

23. For background on West Siberia's initial development and history through 2004, as well as the subsequent decline, see John D. Grace, *Russian Oil Supply: Performance and Prospects,* Oxford Institute for Energy Studies (Oxford: Oxford University Press, 2005), pp. 34–64; Matthew J. Sagers, "Developments in Russian Crude Oil Production in 2000," *Post-Soviet Geography and Economics* 42, no. 3 (April–May 2001): 153–201; Sagers, "West Siberian Oil Production in the Mid-1990s," *International Geology Review* 36, no. 11 (November 1994): 997–1018.

24. The largest part of West Siberian oil production comes from the Khanty-Manssiysk region, with lesser amounts from the Yamal-Nenetsk and Krasnoyarsk regions. Production from Khanty-Manssiysk, the traditional core of the West

Siberian oil province, has been declining, only partially offset by new production from Yamal-Nenetsk and Krasnoyarsk.

25. Gustafson, *Wheel of Fortune,* pp. 460–473.

26. Source: Russian company websites.

27. For the purposes of this book, Rosneft's Vostok Oil complex would be considered "greenfield" because it is primarily onshore and only secondarily "shallow offshore." (We revisit this question below.)

28. A previous version, *Energy Strategy to 2030,* was approved by the Duma without significant debate in 2009. However, it requires regular updating every five years, which would have been 2014. But the update was caught in a bureaucratic impasse as various agencies disagreed with one another, and another five years went by before a new draft version finally appeared in October 2019. The final version of the *Energy Strategy* document was approved by the government in June 2020. The full text is available at http://static.government.ru/media/files/w4sigFOiDjGVDYT4IgsApssm6mZRb7wx.pdf.

29. In late 2020 Novak was elevated to deputy prime minister (one of ten), as described in Chapter 1. At this writing, Pavel Sorokin remains deputy energy minister, and his message is essentially the same as that of the *Energy Strategy.* Novak was replaced by a longtime veteran of the hydropower sector, Nikolai Shulginov; the impact of this change on the positions and roles of the Energy Ministry is not yet clear. It is difficult to tell from a distance what the precise role of Novak and Sorokin may have been in the drafting and debate over the Energy Strategy, but to judge from the prominent role Sorokin has played in defending the strategy in the media, together with the apparently close relationship between Sorokin and Novak, plus the controversy that surrounded its successive versions, I am inclined to believe that the two men played a significant personal role in managing the whole exercise.

30. The *Energy Strategy* document includes major sections on other fossil fuels, as well as analysis of quality issues and the outlook for refined products. But the discussion that follows focuses solely on the outlook for upstream production, given that it is the largest single question determining export revenues.

31. See Tatiana Mitrova et al., *The Future of Oil Production in Russia: Life under Sanctions,* Skolkovo School of Management Energy Program, March 2018, https://energy.skolkovo.ru/downloads/documents/SEneC/research04-en.pdf.

32. Remarks by Pavel Sorokin, speaking at the conference "Innovatsionnaia praktika: Nauka plius bizmes," December 11, 2020, https://ria.ru/20201211/arktika-1588792812.html.

33. Tatiana Mitrova and Vitaly Yermakov, *Russia's Energy Strategy 2035: Struggling to Remain Relevant,* Etudes de l'IFRI, *Russie.Nie.Reports,* no. 28, December 2019, https://www.ifri.org/en/publications/etudes-de-lifri/russias-energy-strategy-2035-struggling-remain-relevant.

34. The source for these numbers is a report on the oil services industry prepared by Deloitte Russia, "Obzor nefteservisnogo rynka Rossii 2019," https://www2.deloitte.com/content/dam/Deloitte/ru/Documents/energy-resources/Russian/oil-gas-russia-survey-2019.pdf.

35. Ibid. The next generation of fracking is multistage fracking, in which several fracks, sometimes twenty or more, are performed along the length of a long-reach horizontal well. It is a very effective technique that produces high initial flow rates, but it is also more expensive and complex. Multistage fracking is rapidly overtaking single-stage. The number of multistage fracs increased from fewer than 100 in 2010 to about 1,000 in 2015 and 2,500 in 2018, and it continues to grow rapidly. In contrast, the number of single-stage fracs peaked in 2014 and is declining, although single-stage operations still remain far more numerous.

36. See "Kak zhivesh', neftiania legenda?," *TEK Rossii,* September 28, 2020, https://www.cdu.ru/tek_russia/articles/1/783/.

37. A chart listing the extent of dependence on foreign sources is included in an April 16, 2019, order (*prikaz* no. 1329) by the minister of industry and trade, Denis Manturov: http://minpromtorg.gov.ru/docs/#!44199.

38. A wide-ranging study of Russian manufacturing, which underscores the general failure to date of import substitution, is V. A. Bazhanov and I. I. Oreshko, "Obrabatyvaiushchie proizvodstva Rossii: Sanktsii, importozameshchenie," *EKO: Vserossiiskii ekonomicheskii zhurnal,* no. 1 (2019): 75–92, available via East View. See also a special issue of *Neftagazovaia vertikal',* April 2021, devoted to the import substitution policy, especially the lead interview with Mikhail Ivaov, the deputy minister of energy in charge of import substitution, http://www.ngv.ru/magazines/article/mikhail-ivanov-my-prodelali-ogromnuyu-rabotu-po-narashchivaniyu-kompetentsiy-rossiyskikh-proizvodite/.

39. Far and away the best treatment of the politics of the collection and distribution of oil rents is the outstanding new book by Adnan Vatansever of King's College London, *Oil in Putin's Russia: The Contests over Rents and Economic Policy* (Toronto: University of Toronto Press, 2021).

40. This conservative bias has come under growing criticism on the grounds that it deprives the economy of funds that could be used for investment. For more on this controversy, see the Conclusion to this book. The best account of the gradual evolution of Vladimir Putin from "saver" to "spender" is Vatansever, *Oil in Putin's Russia.* As Vitaly Yermakov points out, the definition of "lost revenues" by the MinFin is hypocritical: with excessive tax take the tax base would surely shrink, but the calculation assumes that output is perfectly inelastic to rising tax take (private communication with the author).

41. The results of the inventory are reported in "V poiskakh zolotoi serediny," *Neftegazovaia vertikal',* no. 18 (2019), accessed via Factiva at https://professional

.dowjones.com/factiva/. See also the summary given by Orest Kasparov, deputy head of Rosnedra, in the same source. Sorokin's outlook in the fall of 2019, based on the inventory, was even more pessimistic than that of Energy Minister Aleksandr Novak the year before. See Sorokin's remarks at the "Russian Energy Week" conference in October 2019, https://rusenergyweek.com/programme/business -programme-2019/.

42. The results of the inventory are summarized in "V poiskakh zolotoi srediny." See also Sorokin's remarks at the "Russian Energy Week" conference in October 2019, https://rusenergyweek.com/en/news/?PAGEN_1=4.

43. See the extensive quotes in "V poiskakh zolotoi serediny" by Deputy Finance Minister Aleksei Sazanov in defense of the present division of revenues between the industry and the state.

44. Iurii Barsukov and Dmitrii Kozlov, "Dobycha poleznykh oblagaemykh," *Kommersant,* August 3, 2020, https://www.kommersant.ru/doc/4441359?from =main_1. See also the follow-up interview with Rosneft's CEO, Igor Sechin, "MinFin protiv Sechina: Chego zhdat' ot popytki pravitel'stva otniat' u neftianikov 200 mlrd. R," *Znak,* August 3, 2020, https://www.znak.com/2020-08-03/minfin _protiv_sechina_chego_zhdat_ot_popytki_pravitelstva_otnyat_u_neftyanikov _200_mlrd. Part of the Russian response has been to keep the ruble weak, which allows it to balance the budget even when dollar-denominated revenues drop.

45. Private communication to the author. I am grateful to my longtime friend and colleague John Webb for his help in unraveling the tangled politics of the OPEC+ accords.

46. In December 2020 the Vienna Alliance (also known as OPEC+) adjusted its production cuts once more, this time reducing them by 0.5 mbd, effective January 1, 2021. The Saudi government reportedly agreed to take most of the cuts in early 2021. For the latest updates, see the *Financial Times* and, from a Russian perpsective, *Neftegazovaia Vertikal',* http://www.ngv.ru/news/chleny_opek_ne _vypolnili_obyazatelstva_/.

47. In November 2019, on the eve of the pandemic, Igor Sechin wrote an extensive article on the geopolitics of oil in the Russian weekly *Ekspert,* in which he denounced the uncontrolled surge of tight oil in the United States and the US policy, as Sechin viewed it, of acting as a "free rider" by destabilizing global oil markets and harming the interests of established oil producers, chiefly Russia. Sechin's main goal in opposing production cuts, one surmises, was to drive the US tight-oil producers out of business. See "Tsena nestabil'nosti," *Ekspert,* November 2019, https://expert.ru/expert/2019/45/tsena-nestabilnosti/. From 2015 through 2019 Sechin was a frequent commentator in the Russian media on the dangerously unstable state of the global oil market. (*Ekspert* even conveniently featured a dedicated Sechin page, https://expert.ru/avtory/igor_sechin/, that

grouped his articles. However, after late 2019, for whatever reason, Sechin's signed contributions ceased.)

48. Vitalii Petlevoi, "Gosudarstvo budet teriat' na l'gotakh neftianikam do 2.3 trln rublei v god," *Vedomosti*, September 15, 2019, https://www.vedomosti.ru/business/articles/2019/09/15/811275-teryat-lgotah-neftyanikam.

49. US Energy Information Administration, *Technical Recoverable Shale Oil and Gas Resources: Russia* (Washington DC: US Department of Energy, September 2015), https://www.eia.gov/analysis/studies/worldshalegas/pdf/Russia_2013.pdf.

50. Mikhail Silin, "My poka v samom nachale puti osvoeniia Bazhena," *Neftagazovaia vertical'*, no. 13 (2019). In any province there tends to be an inverse relationship between the tight-oil potential and the abundance of mature oil and gas. Thus, in the Middle East the tight-oil potential is thought to be relatively limited. I am indebted for these observations to Keith King, for many years a leading geological expert with Exxon.

51. I am grateful for these insights on tight oil to my friend and former colleague Keith King, who as a petroleum geologist with Exxon brings many years of insight to this question.

52. Quoted in "Decision to Draft Bill on Private Investor Access to Shelf Made during Meeting with Kozak," *Interfax Petroleum Report*, August 22–28, 2019, pp. 5–6.

53. See *Interfax Petroleum*, no. 43, p. 5. The final compromise formula for Vostok Oil—announced, significantly enough, by the minister of finance, Anton Siluanov—grants the Vankor portion of the cluster a discount on the NDPI tax, provided oil prices are above a certain level, but forbids any savings to be used for a pipeline to Payakha and the coast. Rosneft will have to finance the pipeline from its own balance sheet. See "'Rosneft' poslali na Paiiakhu," *Kommersant*, November 8, 2019, https://www.kommersant.ru/doc/4150646.

54. Andrey Chernykh et al., "Burit' na arkticheskom shel'fe ili zhdat'?," *Neftegazovaia Vertikal'*, nos. 3–4 (2019): p. 43.

55. See "Chei golos budet uslyshan? Kto proigraet ot nalogovykh reform v usloviiakh neftianogo kollapsa?," *Neftegazovaia vertikal'*, no. 22 (June 2020), http://www.ngv.ru/magazines/article/chey-golos-budet-uslyshan/?sphrase_id=3041697. This journal has a paywall, but its articles can be accessed via Factiva.

56. In addition, the state grants the oil industry a multitude of special-purpose subsidies and tax breaks, mostly keyed to new oil projects. As a result, it is estimated that the actual tax take is only about three-quarters of the theoretical share. According to the International Monetary Fund, the state's share of the value added from oil and gas extraction and refining stood about 45 percent in 2016, up from 35 percent in 2012. But that is clearly a broader category than oil and gas

exports alone. See Gabriel Di Bella et al., "The Russian State's Size and Its Footprint: Have They Increased?," International Monetary Fund, Working Paper 19 / 53, March 8, 2019, https://www.imf.org/en/Publications/WP/Issues/2019/03/09/The-Russian-States-Size-and-its-Footprint-Have-They-Increased-46662. The share of oil and gas exports alone depends on the exchange rate, but is presumably somewhat less than 39 percent.

57. Another feature of the tax system is that it extracts most of the profit from the brownfields, especially in West Siberia, while directing most of the preferences to greenfield development. The result is that the oil companies derive little cash from the brownfields, and consequently are reluctant to invest in enhanced oil recovery techniques (EOR), for which there are currently no tax preferences. As a result, enhanced recovery techniques remain underused and underdeveloped. According to a recent study by the Skolkovo Energy Center, "virtually no projects" involving EOR applications have been implemented since the imposition of US sanctions in 2014 (even though the US sanctions technically do not apply to them at present), with the exception of a long-standing joint venture between Gazprom Neft and Shell at Salym, which has installed an advanced unit for chemical flooding. See the news report at the website of Salym Petroleum, at https://salympetroleum.com/media/news/spd-started-the-construction-of-asp-mixing-unit/. In January 2020 Shell and Gazprom Neft agreed to expand their joint venture into neighboring regions of Khanty-Mansiisk Okrug. See https://salympetroleum.com/media/news/gazprom-neft-and-shell-to-expand-their-joint-project-to-develop-the-salym-group-of-fields-in-khmao/.

58. For the latest outlook on oil production from the eastern half of the country, see Irina Filimonovo et al., "V ozhidanii vtorogo dykhaniia Vostochno-Sibirskogo regiona," *Neftegazovaia vertikal'*, no. 5 (2021), http://www.ngv.ru/magazines/article/v-ozhidanii-vtorogo-dykhaniya-vostochno-sibirskogo-regiona/. The greenfields tend to be more costly and less productive than those in West Siberia. Even the Vankor cluster, which though located in Krasnoyarsk is geologically part of the West Siberian province, has been disappointing. Vankor, for example, has dropped by half from its peak in 2015, only partly offset by growth from the rest of the Vankor group around it.

59. Normally, according to theory, this should not be the case because Russian greenfield oil will not be the marginal highest-cost oil on the global market. But today's investment decisions for the Russian greenfields are typically taken in the expectation of "normal" oil prices. If actual oil prices end up lower than the ones assumed in the investment decisions, the break-even export price would end up higher than world prices—a result, then, of overoptimism.

60. Darya Korsunskaia and Andrey Ostroukh, "Russia Eyes Budget Surplus for the First Time since 2011," Reuters, https://www.reuters.com/article/us-russia-budget-surplus/russia-eyes-budget-surplus-for-first-time-since-2011-idUSKBN1ICoDS.

3. Can Natural Gas Replace Oil?

1. *Forbes,* https://www.forbes.com/profile/leonid-mikhelson/?sh=2e2663a63140. This estimate dates from December 2020.

2. One of Putin's closest allies, Gennady Timchenko, became a shareholder in 2005 when Novatek conducted an IPO, and he has been a member of the board of directors of Novatek since 2009. Timchenko started out as an oil trader, based in Finland and then Switzerland. He was not initially connected to Putin's Saint Petersburg circle, and his ties to Putin may have originated in the later 2000s through Putin's longtime friend Arkady Rotenberg, because both Timchenko and Rotenberg were involved in pipeline construction. On the origins of Timchenko's involvement in Novatek, see Yelena Mazneva, "Novatek's Largest Shareholder Is Its CEO," *Moscow Times,* December 28, 2010, https://www.themoscowtimes.com/2010/12/28/novateks-largest-shareholder-is-its-ceo-a4057. By 2014, however, Timchenko was considered sufficiently close to Putin to be placed on the US sanctions list.

3. Quoted in Irina Reznik, " Interv'iu: Leonid Mikhel'son, gendirektor i sovladelets kompanii 'Novatek,'" *Vedomosti,* November 2, 2005, http://vspro.media/article/intervyu-leonid-mikhelson-gendirektor-i-sovladelets-kompanii-%C2%ABnovatek%C2%BB. See also Thane Gustafson, *The Bridge: Natural Gas in a Redivided Europe* (Cambridge, MA: Harvard University Press, 2020), chap. 10.

4. See the Novatek website, https://www.novatek.ru/en/about/company/history/.

5. Mikhelson was no stranger to West Siberia. From 1977 to 1985, after graduating as an engineer, he worked on pipeline construction sites in Tiumen' oblast, starting as a foreman. See Wikipedia, https://en.wikipedia.org/wiki/Leonid_Mikhelson, and also Reznik, "Interv'iu."

6. Ivan Igor-Tismenko, "Ne upuskaia moment: Rossiiskii kondensat vospolnit spros na vysokokachestvennuiu neft' v Evrope," *Rusenergy,* June 30, 2005 (author's files). See also Reznik, "Interv'iu."

7. Ielena Mazneva, "Total smenit 'Gazprom' v 'Novateke,'" *Vedomosti,* March 3, 2011, https://www.vedomosti.ru/business/articles/2011/03/03/total_smenit_gazprom_v_novateke.

8. In 2004 Putin attempted to fold Rosneft into Gazprom. This episode is recounted in Thane Gustafson, *Wheel of Fortune: The Battle for Oil and Power in Russia* (Cambridge, MA: Harvard University Press, 2012), pp. 338–351.

9. Mikhail Krutikhin, "Khoziaeva Arktiki: Pochemu Putin i Total poverili v Novatek Mikhel'sona i Timchenko," *Forbes,* May 28, 2018, https://www.forbes.ru/milliardery /362219-hozyaeva-arktiki-pochemu-putin-i-total-poverili-v-novatek-mihelsona-i -timchenko. In January 2018 Mikhelson was included in the US Treasury Department's "CAATSA" list as one of those benefiting from relations with top Russian officials.

10. "Ledovaia obstanovka: Dlia iamal'skogo proekta SPG razrabatyvaiut kontseptsiiu," *Rusenergy,* May 13, 2010; and "Bystro ne poluchitsia: Proektu 'Yamal SPG' ne khvataet gazovykh resursov i inostrannykh partnerov," *Rusenergy,* September 29, 2011 (author's files).

11. Dena Sholk, Mikhail Kuznetsov, and Matthew J. Sagers, *NOVATEK's Yamal LNG: Russian "Mega" Project Remains on Track Despite Challenges* (IHS Markit Strategic Report, January 2017).

12. See Putin's remarks at the November 20, 2019, forum "Russia Is Calling," http://kremlin.ru/events/president/news/62073.

13. Russia has continued to expand its gas reserves steadily. In 2019 it reported 38 trillion cubic meters of proven reserves, up from 34 trillion in 2009. Iran, with 32 trillion, was in second place, followed by Qatar (24.7 trillion) and Turkmenistan (19.5). Source: Statista, https://www.statista.com/statistics/265329/.

14. For a detailed account of this history, see my earlier books, *Crisis amid Plenty* (Princeton, NJ: Princeton University Press, 1989), and *The Bridge.*

15. The first of the Ukraine bypasses, the Yamal-Europe pipeline through Belarus and Poland, was commissioned in 1997. Then, under Putin, came the Blue Stream pipeline to Turkey (2003) and Nord Stream 1 (2012) under the Baltic Sea to Germany. The controversial Nord Stream 2 and Turkstream pipelines are the last of the series.

16. On these points, see the outstanding books and reports of the Russia team at the Oxford Institute for Energy Studies, particularly those of Jonathan Stern, James Henderson, Katya Yafimava, Vitaly Yermakov, and their colleagues. (See oxfordenergy.org/publications.)

17. No one interested in the outlook for Russian gas and gas policy will want to miss James Henderson and Arild Moe's magisterial book *The Globalization of Russian Gas: Political and Commercial Catalysts* (Cheltenham UK: Edward Elgar, 2019), which builds on decades of pathbreaking work by Jonathan Stern and the Oxford Institute of Energy Studies. I am grateful to all three for their kind comments on earlier drafts of this chapter.

18. Calculated based on production statistics for 2018 and 2019 provided by the Ministry of Energy, at https://minenergo.gov.ru/node/1215, and export statistics provided by Russia's Central Bank, at https://www.cbr.ru/statistics/macro_itm/svs/.

19. The one exception is gas supplied by independents to industry. This is sold at unregulated prices.

20. In gas, as in so much else, there are two Russias: in the western third of the country, where most of the population and industry are located, well over half of primary energy consumption comes from gas. In contrast, in the vast eastern two-thirds, only 10 percent of energy consumption is from gas; the rest comes from coal and hydropower. Insofar as most of Russia's future economic growth is likely to occur the western third, where the share of gas is still growing, domestic gas demand in western Russian will probably continue to increase; in the east, the Power of Siberia pipeline may spur regional gas demand, but the overall increase will not be large. Overall, gas demand will remain concentrated in western Russia.

21. This picture is a distillate of the scenarios of leading energy companies and consultancies on the eve of the COVID-19 pandemic.

22. BP, *Energy Outlook, 2020 Edition,* p. 76, https://www.bp.com/content /dam/bp/business-sites/en/global/corporate/pdfs/energy-economics/energy -outlook/bp-energy-outlook-2020.pdf?utm_source=newsletter&utm_medium =email&utm_campaign=newsletter_axiosgenerate&stream=top. BP's 2020 "rapid" scenario differs from its previous year's "rapid transition" scenario (RTS) (pp. 114– 117) in that it assumes a greater role for hydrogen and decarbonization, as well as reflecting "the impact of recent developments" (p. 141). The International Energy Agency (IEA) has likewise grown more cautious. While its base case (called Stated Policies, or STEPS) still sees a 30 percent growth in global gas demand by the 2040s relative to 2019, that growth is entirely concentrated in South and East Asia, while advanced economies show a slight decline. See IEA, *World Energy Outlook 2020,* October 2020, https://www.iea.org/reports/world-energy-outlook-2020.

23. Projection from Rystad, cited in Carbon Tracker, *Decline and Fall: The Size and Vulnerability of the Fossil Fuel System,* https://carbontracker.org/reports/de cline-and-fall/, p. 29.

24. Morgan Stanley, for example, projected a 40 percent decrease in global gas prices by 2030. However, it would be more accurate to say, as for all commodities, "cycles within an overall chronically downward trend."

25. See IEA, *World Energy Investment 2020,* https://www.iea.org/reports/world -energy-investment-2020, pp. 21, 57. At the same time, some LNG suppliers seem undeterred. At this writing, Qatar, significantly, vowed in September 2020 to invest in a massive capacity expansion, after a twelve-year moratorium during which Qatar's competitors threatened to overtake it. See Ben Cahill and Nikos Tsafos, "Qatar's Looming Decisions in LNG Expansion," Commentary from the Center for Strategic and International Studies, September 28, 2020, https://www .csis.org/analysis/qatars-looming-decisions-lng-expansion?utm_source =CSIS+All&utm_campaign=12a33ad981-EMAIL_CAMPAIGN_2018_08_31_06 _36_COPY_01&utm_medium=email&utm_term=0_f326fc46b6-12a33ad981 -222100893.

26. The Oxford Institute of Energy Studies projects capacity of 775 bcm (around 570 Mmta) being reached by 2030, driven by current projects with FID and Qatari expansion. At that rate, 700 Mmta is potentially achievable in the following decade—in other words, slower than previously anticipated, but the overall pre-pandemic target might ultimately still be achieved (James Henderson, private communication to the author).

27. Various scenarios differ on the outlook for coal versus gas in Asia. The IEA is very cautious on the outlook for gas (see its 2019 report, *The Role of Gas in Today's Energy Transitions,* https://www.iea.org/reports/the-role-of-gas-in-todays-energy-transitions, esp. p. 29). In contrast, Morgan Stanley is much more bullish on the prospects for gas in Asia, thanks in particular to lower gas prices. See Morgan Stanley, "Could Liquefied Natural Gas Fuel Global Commodities Disruption?," September 4, 2019, https://www.morganstanley.com/ideas/liquefied-natural-gas.

28. IEA, *The Role of Gas,* p. 28.

29. On the recent turns in Chinese energy policy, and especially coal, see Leslie Hook, "Climate Change: How China Moved from Leader to Laggard," *Financial Times,* November 25, 2019, https://www.ft.com/content/be1250c6-0c4d-11ea-b2d6-9bf4d1957a67; Hook, "China Ramps Up Coal Power in Face of Emissions Efforts," *Financial Times,* November 19, 2019, https://www.ft.com/content/c1feee40-0add-11ea-b2d6-9bf4d1957a67. For background, see Chi-Jen Yang, *Energy Policy in China* (London: Routledge, 2017), chap. 3.

30. The cost advantage of Russian pipeline gas in Europe depends on the accounting, and on the markets. If there is surplus LNG shipping, so that charter rates are at rock bottom, and if there is surplus regasification capacity so that it is running to cover cash costs only, and if tolling arrangements are such that liquefaction costs are low, and if investors in Nord Stream 2 are getting their 20 percent return on equity that they get from Nord Stream 1, then Russia will not be the lowest-cost shipper. I am indebted to Simon Blakey for reminding me of these points, and for his kind and perceptive comments on earlier drafts of this chapter.

31. European Commission, *Committing to Climate-Neutrality by 2050,* https://ec.europa.eu/commission/presscorner/detail/en/ip_20_335.

32. IEA, *The Role of Gas,* p. 64.

33. Climate models do not suggest that "warming" as a global phenomenon inevitably leads to warming of all regional climate zones. If the North Atlantic Oscillation "flips" as a result of climate change (a central thesis of Al Gore's film *An Inconvenient Truth*), then European winters would become harsher.

34. National Energy Administration (NEA), *China Natural Gas Development Report (2019)* (Beijing: Petroleum Industry Press, 2019), p. 6 of the English portion of the report.

35. Ibid., p. 8 of the English portion of the report.

36. Chi-Jen Yang, *Energy Policy in China* (London: Routledge, 2017), chap. 4.

37. NEA, *China Natural Gas Development Report*, p. 9 of the English portion of the report.

38. Global Energy Monitor and the Center for Research on Energy and Clean Air, "A New Coal Boom in China: New Coal Plant Permitting and Proposals Accelerate," June 2020, https://globalenergymonitor.org/wp-content/uploads/2020/06/A-New-Coal-Boom-in-China_English.pdf.

39. Alfred Cang and Jasmine Ng, "China Sets Up National Pipeline Firm in Major Energy Revamp," Bloomberg, December 6, 2019, https://www.bloomberg.com/news/articles/2019-12-06/china-to-launch-major-reform-of-national-oil-pipelines-next-week.

40. Kovykta was discovered in 1987; its companion in Sakha, the Chayanda field, was discovered in 1983. Chayanda is the source of supply for the first phase of "Power of Siberia," Kovykta for the second.

41. "TNK-BP vyhodit iz proekta razrabotki Kovykty," *Kommersant,* May 4, 2010, https://www.kommersant.ru/doc/1364140.

42. For an enthusiastic recap of Russia and China's diplomatic relations, see the speeches of the two countries' heads of state at a recent celebration marking the seventieth anniversary of the relationship. These can be found at http://kremlin.ru/events/president/news/60674.

43. See Mikhail Kalmatskii, "Truba vezet: Chto dast Rossii zapusk gazoprovoda 'Sila Sibiri'," *Izvestiya,* December 2, 2019, https://iz.ru/948866/mikhail-kalmatckii/truba-vezet-chto-dast-rossii-zapusk-gazoprovoda-sila-sibiri.

44. Natal'ia Grib, "Sila 'Gazproma' prirastet Vostokom," *Neftegazovaia vertikal',* no. 20–21 (2020): 19. Alternate ideas include a pipeline from Sakhalin through Vladivostok to China, which has been under discussion for several years, but without visible progress.

45. "Gazprom raz'iasnil vyskazyvanie o peregovorah s Kitaem iz-za koronavirusa," Interfaks, February 13, 2020, https://www.interfax.ru/business/695245.

46. Indeed, some Russian officials are more concerned about the threat to Russian coal exports than they are about gas exports. Aleksandr Ianovskii, a veteran deputy minister of energy whose entire career has been bound up with coal, signed an article in the Energy Ministry's journal, *Energy Policy,* in which he warned against the dangers of Chinese coal. See A. B. Ianovskii, "Na puti v chistoe budushchee: Chto zhdet rossiiskii ugol' v Kitae?," *Energeticheskaia politika,* November 14, 2019, accessed on the website of the Ministry of Energy, https://energypolicy.ru/na-puti-v-chistoe-budushhee-chto-zhdyot-ross/business/2019/21/14/.hy.

47. There had previously been modest shipments of Russian LNG to China from Sakhalin. Grib, "Sila 'Gazproma' prirastet Vostokom," p. 17.

48. NEA, *China Natural Gas Development Report,* p. 9 of the English portion of the report.

49. Ibid. The Russian Energy Ministry's energy strategy to 2035 projects 80 bcma of pipeline capacity to China by 2035.

50. Indian energy companies are also investors in Mozambique LNG, which could become a major supplier to India.

51. See the presentation by Tim Buckley, an authority on Asian energy at the Institute of Energy Economics and Financial Analysis (IEEFA) in Australia. His analysis of the Indian market can be accessed at https://xenetwork.org/ets /episodes/episode-91-energy-transition-in-india-and-southeast-asia-part-1/). See also his regular publications from IEEFA, notably "India: New Record with Renewable Energy Installations 40 Times Higher than Thermal," January 2019, https://ieefa.org/ieefa-india-new-record-with-renewable-energy-installations -40-times-higher-than-thermal/; and "Can India's Ambitious Renewable Energy Goals Be Achieved?," interview with CNBC, March 2020, https://ieefa.org/can -indias-ambitious-renewable-energy-goals-be-achieved/.

52. For an overview of the outlook for petrochemicals (both oil- and gas-based), see I. A. Golubeva et al., "Perspektivy razvitiia neftegazokhimii v Rossii: Proektiruemye i stroiashchiesia neftegazokhimicheskie kompleksy," *Neftegazok-himiia,* no. 2 (2019): 5–12, http://neftegazohimiya.ru/soderzhanie/arkhiv-nomerov -za-2019/neftegazohimiya-2-2019.html; and O. B. Braginskii, "Razvitie otechest-vennoi neftegazokhimii: Korrektirovka kursa," *Neftegazokhimiia,* no. 1 (2019): 5–10, http://neftegazohimiya.ru/soderzhanie/arkhiv-nomerov-za-2019/neftegazohimiya-1 -2019.html.

53. "Plan razvitiia neftgazokhimii: Real'nost' ili blagie pozhelaniia?," *Neftega-zovaia vertikal',* no. 3 (2013): 56–61.

54. Braginskii, "Razvitie otechestvennoi neftegazokhimii," p. 9.

55. Ibid., p. 10. Gas from Chayanda is particularly rich in so-called C2+ fractions, including propane, butane, and ethane. But the rise of domestic Chinese production of these products caused the Chinese Development Bank to cancel its long-term credits to Gazprom for the construction of the Amur cluster, forcing Gazprom to seek more costly short-term credits instead.

56. On the Ust'-Luga complex, see Alina Fadeeva and Petr Kanaev, "Proekt eks-partnera Rotenberga v Ust'-Luge smozhet privlech' den'gi grazhdan," *RBK,* May 25, 2020, https://www.rbc.ru/business/25/05/2020/5ecacobf9a7947bbbbdc3243.

57. Two valuable reports on the potential for Russian exports of methanol and ammonia have been produced by Vygon Consulting in Moscow: *Gazokhimiia*

Rossii, Chast' 1: Metanol: Poka tol'ko plany, March 2019, https://vygon.consulting
/upload/iblock/f22/vygon_consulting_russian_methanol_industry_development
.pdf; and *Gazokhimiia Rossii, Chast' 2: Ammiak: prinimat' s ostorozhnost'iu,*
July 2020, https://vygon.consulting/upload/iblock/eb1/vygon_consulting_ammonia
.pdf.

58. See "Germany Paints Hydrogen Energy Future in Green, Grey, Blue, and
Turquoise," *Hydrogen Processing,* June 11, 2020, https://www.hydrocarbonprocessing
.com/news/2020/06/germany-paints-hydrogen-energy-future-in-green-grey
-blue-and-turquoise.

59. For a review of efforts currently under way in Russia, see A. Mastepanov,
"Vodorodnaia energetika Rossii: Sostoianie i perspektivy," *Energeticheskaia politika* (the monthly journal of the Ministry of Energy) (December 23, 2020),
https://energypolicy.ru/a-mastepanov-vodorodnaya-energetika-rossii
-sostoyanie-i-perspektivy/energetika/2020/14/23/. Hydrogen has been included in
the Ministry of Energy's energy plan to 2035 (discussed in Chapter 1), with a target
of 2 mt of hydrogen exports annually by 2035.

60. See, for example, Tat'iana Mitrova, "Tsena voprosa," *Kommersant,* October 8, 2020, https://www.kommersant.ru/doc/4520083. In December 2020, the
Skolkovo School of Management's Energy Center held a three-day seminar with
counterparts from the EU. One of the major panels was devoted to a joint discussion of hydrogen. According to the host of the event, Yuri Mel'nikov, a government strategy for hydrogen is in preparation, to be released in the second quarter
of 2021. It calls for an initial export of hydrogen of 0.2 mt in 2020. See https://www
.skolkovo.ru/programmes/eu-russia-climate-conference/broadcast/.

61. Gazprom's chief "expert" on hydrogen, Leonid Loginov, has become a frequent contributor in the Russian media, explaining the EU's hydrogen policy, the
debate between partisans of "blue," "turquoise," and "green hydrogen," and the possible long-term implications for Gazprom's European market. Loginov is the
deputy head of Gazprom Export's European Gas Market Monitoring Division. See
the series of articles and interviews by Loginov in *Neftegazovaia vertikal'* in 2018
and 2019, which show an interesting evolution. In 2018 these articles focused primarily on the role of gas as a peak-coverage fuel to offset the rise of power from
renewables. Only in 2019 did Loginov's focus shift to hydrogen. http://www.ngv
.ru/search/?q=%D0%9B%D0%BE%D0%B3%D0%B8%D0%BD%D0%BE%D0%
B2. See, in particular, "Stavka na vodorod," September 18, 2019, http://www.ngv
.ru/magazines/article/stavka-na-vodorod/?sphrase_id=2319206.

62. One of the first articles in the mass media to discuss this issue is Kirill
Astakhov, "Preobrazovanie izlishkov elektrichestva v metan i vodorod," *Nezavisimaia gazeta,* January 15, 2019, p. 14, https://www.ng.ru/ng_energiya/2019-01-14/14
_7481_energy07.html.

63. There is currently a great deal of experimentation, debate, and regulatory speculation regarding "blending limits." Everyone agrees that 2 percent is acceptable, although Dutch authorities put the limit at 9 percent. It is important where the blended methane-hydrogen is going to be used. If the grid that it is injected into feeds compressed natural gas into filling stations, for example, then more than 2 percent is not safe. And if the part of the grid's system involves certain types of storage facility, then underground migration and leakage pose serious problems. At present it does not seem likely that more than 5 percent hydrogen could be shipped safely through a large-diameter pipeline.

64. Gazprom press release, September 2019, https://www.gazprom.ru/press/news/2019/september/article488231/.

65. André Ballin, "Russlands neue Energiestrategie: Nord Stream 2 soll Wasserstoff Liefern," *Handelsblatt,* July 27, 2020, https://www.handelsblatt.com/politik/international/gazprom-und-rosatom-russlands-neue-energiestrategie-nord-stream-2-soll-wasserstoff-liefern/26039724.html. Several turbine manufacturers already advertise that they have developed "hydrogen-ready" turbines.

66. "Otkaz ot gaza destabiliziruet ekonomiku Evropy," *Neftegazovaia vertikal',* November 6, 2019, http://www.ngv.ru/news/otkaz_ot_gaza_destabiliziruet_ekonomiku_evropy/?sphrase_id=2318768.

67. For current developments in Mozambique LNG, which have caused Total to suspend its work there, see https://www.mzlng.total.com/en/about-mozambique-liquefied-natural-gas-project.

4. Russia's Coal Dilemma

1. "Russian Authorities Cover Snow in White Paint to Hide Signs of Pollution, Reports," *Moscow Times,* December 19, 2018, https://www.themoscowtimes.com/2018/12/19/russian-authorities-cover-snow-white-paint-hide-sings-pollution-a63892.

2. The scene was recorded on video; see https://www.dailymail.co.uk/video/news/video-1657416/Video-Russian-official-kneels-crowd-ask-forgiveness-fire.html.

3. "V Kuzbasse 'prostaivaet' 23 mln t eksportnogo uglia," dprom.online, https://dprom.online/chindustry/v-kuzbasse-prostaivaet-23-mln-t-eksportnogo-uglya/. The governor's daily blog is available through multiple social-media platforms, including vKontakte and Telegram. See the YouTube version at https://www.youtube.com/channel/UCB26kmhyN5dD-K2ZWJ2QInw. See the presidential website, http://en.kremlin.ru/events/president/news/56943.

4. BP, *Statistical Review 2020,* https://www.bp.com/content/dam/bp/business-sites/en/global/corporate/pdfs/energy-economics/statistical-review/bp-stats

-review-2020-co2-emissions.pdf. Energy, in turn, produces about three-quarters of all greenhouse gases worldwide.

5. International Energy Agency, *Global Energy and CO₂ Status Report 2019,* https://www.iea.org/reports/global-energy-co2-status-report-2019/emissions.

6. BP, *Statistical Review of World Energy 2020,* https://www.bp.com/en/global/corporate/energy-economics/statistical-review-of-world-energy/primary-energy.html.

7. E. T. Gaidar Institute of Economic Policy, *Ekonomicheskaia politika Rossii: Turbulentnoe desiatiletie, 2008–2018* (Moscow: Delo, 2020), https://www.iep.ru/files/text/trends/2019/2019.pdf. I am grateful to Dr. Irina Dezhina for alerting me to this document, which contains valuable chapters on all the major sectors of the Russian economy.

8. Federal Customs Service, report for 2019, accessed at https://customs.gov.ru/statistic/vneshn-torg/vneshn-torg-countries.

9. See Bloomberg New Energy Finance (BNEF), *New Energy Outlook 2020.* See the online presentation by BNEF's chief economist, Seb Henbest, at the Center for Strategic and International Studies, December 4, 2020, https://www.csis.org/events/online-event-bloombergnefs-new-energy-outlook-2020.

10. International Energy Agency, *2019 CO₂ Status Report,* https://www.iea.org/reports/global-energy-co2-status-report-2019/emissions.

11. Domestic prices for natural gas have risen substantially in recent years, but the Russian government has retreated from its previously announced policy of raising them to "export netback" levels. As a result, natural gas prices remain effectively subsidized at low levels in the domestic market, preventing coal from competing with natural gas except in the eastern part of the country.

12. A useful review of the state of coal-fired generation in Russia and the outlook to 2040 is Aleksei Khokhlov and Iurii Mel'nikov, *Ugol'naia generatsiia: Novye vyzovy i vozmozhnosti* (Moscow: Skolkovo Energy Center, January 2019), https://energy.skolkovo.ru/downloads/documents/SEneC/Research/SKOLKOVO_EneC_Coal_generation_2019.01.01_Rus.pdf.

13. International Energy Agency, "Asia Is Set to Support Global Coal Demand for the Next Five Years," December 17, 2019, at https://www.iea.org/news/asia-is-set-to-support-global-coal-demand-for-the-next-five-years.

14. For a report on subsidies for coal production and the construction of coal-fired power plants, see Overseas Development Institute (ODI), *G20 Coal Subsidies: Tracking Government Support to a Fading Industry,* June 2019, https://www.odi.org/publications/11355-g20-coal-subsidies-tracking-government-support-fading-industry.

15. "Evropa otkazyvaetsia ot uglia, no Rossiia ne verit v poteriu rynka," DW (the online service of Deutsche Welle), September 18, 2019, https://www.dw.com/ru

/европа-отказывается-от-угля-но-россия-не-верит-в-потерю-рынка/a
-50462504.

16. *Kohleausstieggesetz* (Law on the Exit from Coal), http://www.gesetze-im
-internet.de/kohleausg/.

17. Kathrin Witsch, "Deutsche Konzerne geraten wegen Kohleimporten aus
Russland in Erklärungsnot," *Handelsblatt,* November 4, 2019. This may work to
the temporary advantage of Russian gas, as discussed in Chapter 3, as Germany
deals with the short-term consequences of its impending exit from nuclear.

18. International Energy Agency, *Coal 2019: Analysis and Forecasts to 2024,* December 2019, https://www.iea.org/reports/coal-2019.

19. See, for instance, Russian Ministry of Energy, "Program for Developing the
Coal Industry until the Year 2035," June 13, 2020, at https://minenergo.gov.ru
/node/433.

20. Carlos Fernández Alvarez, "Fading Fast in the US and Europe, Coal Still
Reigns in Asia," International Energy Agency, December 12, 2019, https://www
.iea.org/commentaries/fading-fast-in-the-us-and-europe-coal-still-reigns-in
-asia.

21. Catherine Shearer, "A New Coal Boom in China," Global Energy Monitor
Briefing, June 2020, https://globalenergymonitor.org/wp-content/uploads/2020
/06/China-coal-plant-brief-June-2020v2.pdf.

22. For a comprehensive update on the latest developments in Chinese climate
and energy policy, see Lauri Myllyvirta et al., "Political Economy of Climate and
Clean Energy in China," https://www.boell.de/en/2021/01/12/political-economy
-climate-and-clean-energy-china.

23. Ibid.

24. Presentation on India by Australian analyst Tim Buckley, *Energy Transition
Show,* episode 91, March 20, 2019.

25. ODI Report, *G20 Coal Subsidies: Tracking Government Support to a
Fading Industry, 2019.* Japan is also a major investor in coal-fired power outside
its borders, with financing from the Japan Bank for International Cooperation and
strong support from Japanese equipment manufacturers. In 2020, however, the
Japanese government announced it is ending this policy (*Financial Times,* July 14,
2020).

26. Bloomberg New Energy Finance, "Scale-Up of Solar and Wind Puts Existing Coal, Gas, at Risk," April 28, 2020, https://about.bnef.com/blog/scale-up-of
-solar-and-wind-puts-existing-coal-gas-at-risk/?sf121491850=1.

27. Carnegie Endowment for International Peace, *The Formation and Evolution
of the Soviet Union's Oil and Gas Dependence,* https://carnegieendowment.org/2017
/03/29/formation-and-evolution-of-soviet-union-s-oil-and-gas-dependence-pub
-68443#_edn1.

28. For a survey of the state of the Russian coal industry at the end of the 1980s, see Albina Tretyakova and Matthew J. Sagers, *Production Costs and Subsidies in the Soviet Coal Industry* (Washington, DC: US Bureau of the Census, 1988).

29. For a detailed history of Russian coal, see the portal of Rosugol''s coal museum, https://www.rosugol.ru/museum/?PAGEN_1=5, as well as former energy minister Alexander Novak's column, found at https://minenergo.gov.ru/node /18573.

30. https://rosstat.gov.ru/storage/mediabank/Ejegodnik_2019.pdf, p. 390. Of this, 79 percent (or 92 mt) consisted of steam coal, mainly intended for the Russian Far East and for export; 21 percent (or 92 mt) consisted of coking coal ("metallurgical coal"), mainly for the steel industry.

31. *Russian Energy Strategy to 2035,* p. 18, http://static.government.ru/media/files /w4sigFOiDjGVDYT4IgsApssm6mZRb7wx.pdf.

32. L. S. Plakitkina and Iu. A. Plakitkin, "Novye stsenarii razvitiia ekonomiki Rossii: aktualizirovannye prognozy razvitiia dobychi uglia v period do 2025g," *Ugol'* (May 2018): 66–71, https://www.eriras.ru/files/plakitkina_-17-_prognoz_dobychi _zh.ugol_03.05.2018_-shchshch1-.pdf.

33. The exact figure in 2018 was 54 percent. See the website of the Ministry of Energy, https://minenergo.gov.ru/node/437.

34. Galina Muzlova, "Novye vozmozhnosti dlia eksporta uglia," *Morskie vesti Rossii,* August 8, 2019, http://www.morvesti.ru/analitika/1688/79974/.

35. A good case study of the state of coal-fired generation is Khokhlov and Mel'nikov, *Ugol'naia generatsiia.*

36. Ibid., pp. 59–62.

37. See the detailed listing under the heading "Restructuring of the Coal Sector" on the website of the Ministry of Energy, https://minenergo.gov.ru/node/438.

38. "Na slome trendov: Kakoe buduschee zhdet rossiiskii ugol'nyi eksport," PWC, 2020, https://www.pwc.ru/ru/industries/mining-and-metals/na-slome -trendov-kakoe-budushchee-zhdet-rossijskij-ugolnyj-eksport.pdf. Between 2008 and 2018, capital investment in the coal industry, mostly from private sources, grew from 60 billion rubles in 2008 (then about $2.6 billion) to 144 billion rubles (then still only about $2.4 billion, owing to intervening devaluations of the ruble) in 2018. In other words, in dollar terms, coal investment hardly budged; but because most of the inputs into the industry are denominated in rubles, the nominal increase does reflect a real growth in inputs. (The figures on capital spending in the coal industry appear in the government's report *Russian Energy Strategy to 2035,* p. 18.)

39. "Vostochnyi poligon poluchil zelieyi svet," RBC, February 10, 2020, https://www.rbc.ru/newspaper/2020/02/10/5e3d83999a794763c6d0d675.

40. "Kak rasshirit' 'butylochnoe gorlyshko'?," *TEK Rossii,* no. 10 (2019), http://www.cdu.ru/tek_russia/articles/5/641/.

41. RZhD website, https://company.rzd.ru/ru/9423. The sources of RZhD's investment are described as RZhD's own internal resources, plus funds from state and regional budgets. According to the company's 2019 report, around 90 percent of its total annual investment comes from internal resources; https://company.rzd.ru/ru/9471#Annual. The annual report puts the Eastern Polygon at the top of RZhD's investment priorities (p. 58). The Eastern Polygon was created in 2016, from a merger of four eastern railroad systems, Krasnoyarskaia, Vostochno-Sibirskaia, Zabaikal'skaia, and Dal'nevostochnaia (p. 316). Estimating the equivalents in dollar amounts is complicated by the fact that in mid-2015 the Russian Central Bank sharply devalued the ruble. I have taken an average for that year of 60 rubles to the dollar.

42. Danila Bochkarev, "Uglerodnyi nalog ES i strategii rossiiskikh kompanii," *Gazeta.ru*, November 25, 2020, https://www.gazeta.ru/comments/2020/11/25_a_13373707.shtml.

43. See the discussion by Ellie Martus, "Russian Industry Responses to Climate Change: The Case of the Metals and Mining Sector," *Climate Policy* 19, no. 1 (2018): 17–29.

44. SUEK's new director general, Stepan Solzhenitsyn (a son of noted author Aleksandr Solzhenitsyn), is a striking example. His appointment in May 2020 surprised the Russian energy world. It is in many ways a remarkable appointment. Solzhenitsyn is a graduate of both MIT and Harvard and worked as a top energy analyst for McKinsey Russia from 2004 to 2018, reaching the rank of partner (*Novye izvestiia*, May 19, 2020).

45. With a French engineering degree and an MBA from Wharton Business School, Tuzov spent much of his early career abroad, including a lengthy stint with BCG (formerly known as the Boston Consulting Group).

46. The next two paragraphs are a summary of a presentation by Vladimir Tuzov at the July 2019 Skolkovo Energy Institute Summer School, https://www.youtube.com/watch?v=-MOVnHC4DEo.

47. A. B. Ianovskii, "Na puti v chistoe budushchee: Chto zhdet rossiiskii ugol' v Kitae?," *Energeticheskaia politika*, November 14, 2019, https://energypolicy.ru/?p=2407.

48. Tuzov, July 2019 presentation, minute 24. Russian coal exports declined the following year, to 208 mt in 2019. See the annual reports of the Russian Customs Service, https://customs.gov.ru/statistic/%D0%AF%D0%BD%D0%B2%D0%B0%D1%80%D1%8C%20-%20%D0%BC%D0%B0%D0%B9%202018.

49. *Russian Energy Strategy to 2035*, p. 19.

50. Transcript of interview with Tat'iana Lan'shina, "Ugol' mertv. Nuzhno rassmatrivat' stsenarii, v kotorykh mir rezko sokratit potreblenie iskopaemogo topliva," *Respublika (Slon)*, March 28, 2020, accessed via East View (https://dlib.east

view.com/). Lam'shina is an economist at the Russian Academy on the Economy and Public Administration (RANEPA) in Moscow. As this quote suggests, she is a strong voice in favor of a more proactive Russian policy toward climate change.

51. *Rossiiskii Statisticheskii Ezhegodnik 2019,* section 26.18.

52. In 2019 Russia's exports to China were 33 mt, compared to 63 mt to the European Union. For Russian coal exports by country of destination, see Statista, https://www.statista.com/statistics/1066718/russian-coal-export-volume-by -destination/#statisticContainer. The original source is the Russian Customs Service, referenced in note 48.

5. Renewables

1. See, for instance, Lev Lur'ie, "Zhizn' i sud'ba otsa rossiiskoi privatizatsii," *Delovoi Petersburg,* July 11, 2015, https://www.dp.ru/a/2015/07/10/CHernaja_metka _Cubajsa.

2. Andrei Kolesnikov, "Poliubite SPS ryzhen'koi," Gazeta.ru, July 29, 2003, https://www.gazeta.ru/column/kolesnikov/171255.shtml.

3. Chubais's own rationale and battle plan for the UES restructuring, shortly after Putin was first elected president in 2000, were laid out in his landmark speech at the All-Russian conference on the problems of the fuel and power complex, "Izlozhenie vystupleniia A.B. Chubaisa," Surgut, March 2000. (Chubais was originally trained as an electrical engineer at the Moscow Power Institute.)

4. Rusnano was created in 2007. The following year, its mission ended by Chubais's restructuring, UES went out of existence.

5. In September 2019 Chubais launched a regular series of podcasts, of which the first three were devoted to the future of renewables in Russia. The entire series can still be read and heard on the website of Rusnano, at https://www.rusnano .com/about/spetsproekty/podkastnanosvod. It is also available on various social media, such as Vkontakte, SoundCloud, and iTunes. Chubais also appears frequently as a lecturer in various forums, such as his presentation at the Bauman Moscow State Technical University, described at the beginning of this chapter. See "Vozobnovliaemaia energetika v Rossii," October 14, 2019, https://www.rusnano .com/about/press-centre/first-person/podkast-nanosvod.

6. "Putin naznachil Chubaisa na novuiu dolzhnost," RBC, December 4, 2020, https://www.rbc.ru/politics/04/12/2020/5fca72019a79470a666599d5. It is not yet clear at this writing what Chubais's precise duties may be, or whether his new appointment will turn out to be little more than an honorable retirement. Rusnano, meanwhile, has been folded into VEB.RF, a state-owned investment bank devoted to promoting high-tech innovation.

7. Chubais, "Vozobnovliaemaia energetika v Rossii."

8. Chubais said "three-quarters," but that appears to be a mistake. See https://www.rosteplo.ru/Tech_stat/stat_shablon.php?id=283.

9. See the website of the Ministry of Energy, https://minenergo.gov.ru/node /489.

10. Transcript on the presidential website, http://kremlin.ru/events/president /news/60109. The reorganization was mentioned in note 6. See the website of Rusnano, https://en.rusnano.com/about.

11. Bloomberg New Energy Finance, *New Energy Outlook 2019,* https://www .gihub.org/resources/publications/bnef-new-energy-outlook-2019/. See also the annual report of the International Renewable Energy Agency, *Renewable Power Generation Costs in 2019,* https://www.irena.org/-/media/Files/IRENA/Agency /Publication/2020/Jun/IRENA_Power_Generation_Costs_2019.pdf.

12. This is in terms of current operating expenses. On a "levelized cost of electricity basis" (LCOE), wind and power have not yet overtaken gas-fired generation; that is, on the basis of lifetime costs, it is still cheaper to build a new gas-fired plant than a new utility-scale solar or a wind farm. But given current cost trends, that advantage is rapidly disappearing.

13. By comparison, global investment in oil and gas (both upstream and downstream) was about $710 billion in the same year. International Energy Agency, *World Energy Investment 2019* (Paris: International Energy Agency, 2020), https://www .iea.org/reports/world-energy-investment-2019/introduction#abstract.

14. Bloomberg New Energy Finance, *Clean Energy Investment Trends 2019,* https://data.bloomberglp.com/professional/sites/24/BloombergNEF-Clean -Energy-Investment-Trends-2019.pdf. Bloomberg defines "clean energy" very broadly, which is confusing. Thus the $363 billion number it gives as the clean energy total for 2019 includes, for example, Tesla's IPO. Nevertheless, the overall trend is clear.

15. See, for example, EY's "Renewable Revolution" scenario, *Energy Reimagined,* 2018, https://www.ey.com/en_us/energy-reimagined/what-is-the-recipe-for -tomorrows-energy-mix; and International Renewable Energy Agency, *Global Renewables Outlook: Energy Transformation 2050,* "Planned Energy Scenario" (April 2020), p. 25, fig. S.3, https://www.irena.org/publications/2020/Apr/Global -Renewables-Outlook-2020.

16. The most up-to-date set of scenarios at this writing comes from Bloomberg New Energy Finance (BNEF), a consultancy that attempts to provide a balanced view based on financial analysis. According to BNEF's chief economist, Seb Henbest, presenting BNEF's "New Energy Outlook 2020" at the Center for Strategic and International Studies (CSIS) in December 2020, renewables will account for 69 percent of electricity generation worldwide by 2050, while fossil fuels will

provide only 24 percent. See Henbest's online presentation at https://www.youtube .com/watch?v=toDsii7T-as, slide 8. If nuclear is included among the renewables, the share of renewables grows to 76 percent.

17. I am grateful to one of my students at Georgetown University, Matias Burdman, for his enlightening paper on battery technology, written for my seminar "Energy and the Coming World." The paper is available from Matias Burdman.

18. On the difficulties of building long-distance power transmission systems to link renewable power to centers of demand, see the thought-provoking account by Russell Gold, *Superpower: One's Man Quest to Transform American Energy* (New York: Simon and Schuster, 2019).

19. Henbest's CSIS presentation, slide 14.

20. Not surprisingly, Gazprom, which owns the largest single share of generating capacity in Russia through its subsidiary Energoholding, is strongly opposed to the development of renewables on the grounds that it competes for scarce funds with the more important strategic task of modernizing Russia's gas-fired fleet. For a strong statement of skepticism, see the extended interview with the head of Gazprom Energoholding, Denis Fedorov, in "Samymi krupnymi nepla tel'shchikami iavliaiutsia bogatye liudi," *Kommersant,* February 3, 2021, https:// www.kommersant.ru/doc/4672421. So far, Gazprom's only significant involvement in renewables is through hydropower.

21. Sergei Golubchikov, "Energetika Severa: problem i puti ikh resheniia," *Energiia,* no. 11, 2002, https://www.rosteplo.ru/Tech_stat/stat_shablon.php?id=283.

22. Only 56 percent of all projects begun since 2014 have met their construction targets. There are widespread delays, for which the investors are required to pay fines. The main reason was the economic crisis of 2014–2016 and domestic suppliers' failure to perform (*TEK Rossii,* no. 3 [2020]: 10ff.). See also Polina Smertina, "Zelenaia energetika zaderzhivaetsia," *Kommersant,* February 19, 2020, https://www .kommersant.ru/doc/4260079. The article is based on a report by an agency called the Center for Financial Accounts, which is part of the Market Council that oversees the power industry.

23. Website of the Russian Ministry of Energy, https://minenergo.gov.ru /node/532.

24. For Novak's report to Putin, see http://kremlin.ru/events/president/news /59660.

25. For a useful checklist of the efforts undertaken by Russian oil and gas companies as of the end of 2020, see Vladislav Karasevich, "Puti dekarbonizatsii v rossiikikh neftegazovykh kompanii," *Neftegazovaia vertikal',* no. 6 (2021), http://www.ngv.ru/magazines/article/puti-dekarbonizatsii-v-rossiyskikh -neftegazovykh-kompaniyakh/. Karasevich heads a department (*kafedra*) of renewable energy at Russia's prestigious Gubkin Institute of Oil and Gas, and is

thus well placed to follow the activities of the oil and gas companies in the renewables sector.

26. For a history of the restructuring of the Russian power industry in the 2000s, see Susanne A. Wengle, *Post-Soviet Power: State-Led Development and Russia's Marketization* (Cambridge: Cambridge University Press, 2015). In Siberia, where the power sector was largely owned by the aluminum industry and low-cost hydropower was passed on to the smelters at super-low prices, the very concept of a market, whether wholesale or otherwise, was largely meaningless. But in the two western zones, politics were more favorable to market-oriented reforms.

27. Speech by Anatoly Chubais at the Bauman Moscow State Technical University, October 4, 2019, https://www.rusnano.com/about/press-centre/first-person/podkast-nanosvod.

28. Enel Russia buys its wind turbines from Siemens Gamesa. As of early 2020 it had three wind farms under construction, two in south Russia (scheduled for completion in 2020) and one on the Kola Peninsula (scheduled for completion in 2021). See the company's website at https://www.enelrussia.ru/en/about-us.html.

29. Ibid.

30. From slides presented by Chubais in his Bauman speech.

31. "Russia Considers Banning Foreign Companies from Renewable Energy Projects," *Moscow Times,* October 16, 2019, https://www.themoscowtimes.com/2019/10/16/russia-considers-banning-foreign-companies-from-renewable-energy-projects-a67755.

32. Renova had actually been created in 1990 as a Russian-American joint venture. For more history on Vekselberg and Renova, see "Kak menialis' aktivy Vekselberga i pri chiom zdes' Medvedev," *Regnum,* December 3, 2019, https://regnum.ru/news/economy/2796106.html.

33. Anna Peretolchina, "Novoe 'solntse' Chubaisa," *Vedomosti,* June 5, 2009, accessed via East View, https://dlib.eastview.com/.

34. According to another source, Rusnano in 2009 invested $400 million in the new company.

35. Oerlikon Solar, a subsidiary of one of the most famous names in Swiss manufacturing, was founded in 2006. It had a brief but tumultuous history. In that same year Renova purchased a 10.3 percent stake in its parent company, Oerlikon, and in subsequent years Renova played an increasing role in its management. At the time it was founded, Oerlikon Solar was the world's only provider of turnkey factories for the mass production of large-surface thin-film solar modules. Initially highly profitable, Oerlikon Solar fell victim to the surge of solar-panel production in China, which soon drove competitors out of business worldwide. By 2012 Oerlikon Solar was losing money, and in that year it was sold to Tokyo Electron Limited and was renamed TEL Solar. By 2014 TEL Solar was out of business.

36. Andrei Zuev, "Novaia energetika Rossii," *TEK Rossii,* no. 3 (2020): 22.

37. Rusnano has exited from direct involvement in Hevel; in 2018 it sold off its 49 percent share to a group called Ream Management, which appears to be affiliated with Renova. Renova, meanwhile, has also scaled down its stake in Hevel, but remains active in the management. The present CEO of Hevel is Igor Shakhray. He joined Renova in 2007 and Hevel in 2010. He has been CEO since 2015. His predecessor, Igor Akhmerov, had been the CFO of Renova before becoming the CEO of Hevel in 2012–2015. From the beginning, the top management of Hevel has consisted of senior executives from Renova, as it does to the present day.

38. "The Rise of China's Solar Industry: How Bright Is Its Future?," Knowledge @ Wharton, March 26, 2008, https://knowledge.wharton.upenn.edu/article/the-rise-of-chinas-solar-industry-how-bright-is-its-future/.

39. On the abrupt rise of the Chinese solar industry and the resulting upheaval in the global solar industry, see Varun Sivaram, *Taming the Sun* (Cambridge, MA: MIT Press, 2018), chap. 2.

40. The very first Russian producer of silicon wafers, located in Mytishchi (Moscow Oblast) was Gelios-Resurs, which ironically was exporting to Germany, but it stopped exporting when the German government cut its subsidies for solar. It refocused its business on Russia, with a planned plant in Mordoviia. Gelios-Resurs was a private venture, largely financed by a Russian entrepreneur, Aleksandr Ors. "Na zakate Evropy," *Kommersant,* December 3, 2013.

41. Hevel website, https://www.hevelsolar.com/en/about/.

42. Solar Energy Industries Association, "US Solar Market Insight," September 10, 2020, https://www.seia.org/us-solar-market-insight#:~:text=The%20 U.S.%20installed%203.5%20gigawatts,power%2016.1%20million%20American %20homes.

43. See the website of Solar Systems at http://en.solarsystems.msk.ru/. Solar Systems is described as a subsidiary of Amur Sirius, itself a subsidiary of Harbin Electric. It consists of a mixture of Chinese and Russian players; there are two Chinese members on its board of directors, and the China Construction Bank is a shareholder. Two of its partners, Hopewind and KSTAR, are based in Guangdong.

44. Renewables Now, *Amur Sirius Unit to Build 100-MW-Plus Solar Panel Factory in Russia,* https://renewablesnow.com/news/amur-sirius-unit-to-build-100-mw-plus-solar-panel-factory-in-russia-report-445742/.

45. Website for Solar Silicon Technologies, LLC, https://www.zawya.com/mena/en/company/Solar_Silicon_Technologies_LLC-181212143351/. Chubais's comment appears in his 2019 speech at Bauman Moscow State Technical University.

46. Zuev, "Novaia energetika Rossii," 10ff., and Karasevich, "Puti dekarbonizatsii."

47. Andrei Zuev, "Al'ternativnaia energetika," *TEK Rossii,* no. 2 (February 2019): 13–25. See also news item in *TEK Rossii,* no. 6 (2019): 4. For an update, see Zuev, "Novaia energetika Rossii," 21–23.

48. Zuev, "Novaia energetika Rossii," 21.

49. Rosatom's chief trainer in Germany is Enercon in Magdeburg, the biggest producer of windmills in Germany. See Kirill Astakhov, "Nastroika Adygeiskoi vetroelektrostantsii," *Nezavisimaia gazeta,* December 10, 2019, p. 16.

50. Rosatom annual reports, https://rosatom.ru/en/about-us/public-reporting/ and https://www.rosatom.ru/upload/iblock/oba/oba23d180bc202e22b53b62ca57a 25bb.pdf.

51. Indeed, Rosatom as a whole is no particular friend of renewables. In 2017 deputy CEO Kirill Komarov was quoted as saying, "To provide the same power as a 1,200MW nuclear plant with a footprint of 1 sq km, you would need a wind farm the size of Andorra and a solar panel the size of Copenhagen," adding that "the answer is in combination" (Henry Foy, "Rosatom Powers through Nuclear Industry Woes," *Financial Times,* June 28, 2017, https://www.ft.com/content/774358b4-5a4a -11e7-9bc8-8055f264aa8b).

52. The first of the Stavropol windmills, the Kochubeievskaia, began generating power in January 2021.

53. Astakhov, "Nastroika Adygeiskoi vetroelektrostantsii." This assumes an exchange rate of 65 rubles per US dollar, in 2018 dollars. See also the website of NovaWind, http://novawind.ru/production/partners/.

54. Zuev, "Novaia energetika Rossii," 21.

55. The remaining shareholders consist of a mix of investment funds and smaller minority shareholders. The company is listed on the Moscow Stock Exchange (MICEX).

56. Nikiforov, "Vazhnyi shag." Enel has three wind projects in Russia for a total capacity as of June 2019 of 362 MW, of which 90 is under construction. In Rostov Oblast, Enel built the Azovskaya wind park, which was commissioned in 2021. It is supported with funding from the European Development Bank. In addition, Enel has two more wind projects, one in Stavropol and the other on the Kola Peninsula. The latter is the only Russian wind project in the Far North. See also the Enel company website, https://www.enelrussia.ru/en.html.

57. Kirill Astakhov, "Potentsial perekhoda na vozobnobliaemye resursy," *Nezavisimaia gazeta,* June 19, 2018, p. 12. In his meeting with Putin in March 2019, Chubais said six universities.

58. Nevertheless, the project was approved by Minpromtorg as being 65 percent based on Russian technology. This allows the wind park to receive payment. So Minpromtorg is evidently the watchdog on the import substitution (*lokalizatsiia*) policy. See Rusnano website, https://www.rusnano.com/about/press

-centre/news/20200302-rosnano-elektroenergiya-pervogo-vetroparka-v-rostovskoy
-oblasti-postupila-na-optoviy-rynok.

59. Website of the Association for Renewable Energy, https://rreda.ru/.

60. Astakhov, "Potentsial perekhoda." This is evidently different from Oerlikon Solar and is instead a part of the parent company that produces a wide range of manufactured products and remains a Renova property.

61. It employs 140 people. The project cost 800 million rubles, of which Rusnano contributed 196 million. Windar Renovables (a Spanish company, part of the Daniel Alonso Group) and Severstal are minority shareholders. See https://www .rusnano.com/projects/portfolio/bashni-vrs.

62. Rusnano website, https://www.rusnano.com/about/press-centre/news/2020 0409-rosnano-vestas-predostavil-komplekty-sredstv-zashchity-dlya-peri natalnogo-tsentra-v-ulyanovske.

63. Lev Moskovkin, "Ne nado idealizirovat' Chubaisa, on sam spravitsia," *Moskovskaia pravda*, March 25, 2019, p. 7.

64. Oleg Nikiforov, "Vazhnyi shag k realizatsii investprogrammy RF po vozobnovliaemoi energetiki," *Nezavisimaia gazeta*, November 12, 2019, accessed via East View.

65. US Department of Energy, Office of Energy Efficiency and Renewable Energy, *2018 Offshore Wind Technologies Market Report*, https://www.energy.gov /sites/prod/files/2019/08/f65/2018%20Offshore%20Wind%20Market%20Report .pdf, p. 23.

66. International Renewable Energy Agency, *Future of Wind*, October 2019, https://www.irena.org/-/media/Files/IRENA/Agency/Publication/2019/Oct /IRENA_Future_of_wind_2019.pdf.

67. International Energy Agency, *Offshore Wind Outlook 2019*, https://www.iea .org/reports/offshore-wind-outlook-2019.

68. Ibid.

69. The website of the Ministry of Energy gives regular numbers on renewables output and capacity. See https://minenergo.gov.ru/node/532.

70. Chubais's Bauman video, minute 56.

71. "Vlasti khotiat zapretit' inostrantsam proektirovat' v Rossii 'zelenye' elektrostantsii," *Vedomosti*, October 15, 2019. The Ministry of Industry and Trade has also taken a hard line on the participation of foreign companies and ships in the Russian Arctic.

6. *The Revival of Russian Nuclear Power*

1. For a description of the origins of the Russian debt crisis and the government's default in 1998, see Thane Gustafson, *Capitalism Russian-Style* (Cambridge:

Cambridge University Press, 1998), esp. prologue, pp. 1–9. On Yeltsin's search for a successor, see Timothy J. Colton's magisterial biography *Yeltsin: A Life* (New York: Basic Books, 2011).

2. Krasnoye Sormovo, one of Russia's oldest factories with origins going back to prerevolutionary times, built nuclear submarines, along with a wide range of civilian products, including ships for river traffic but also agricultural machinery and even washing machines. Kirienko, who had graduated from the Institute of Water Transportation, where his father was a professor, was mainly active in the plant's Komsomol organization.

3. See the profiles of the Krasnoye Sormovo plant and of Sergei Kirienko at http://krsormovo.nnov.ru and http://www.kremlin.ru/catalog/persons/175/biography, respectively.

4. Putin's press conference on November 22, 2005, quoted in Aleksei Nikol'skii and Tat'iana Egorova, "Kirienko razdelit Rosatom," *Vedomosti,* November 23, 2005, accessed via East View (https://dlib.eastview.com/).

5. International Energy Agency (IEA), *Nuclear Power in a Clean Energy System* (Paris: IEA, May 2019), www.iea.org/publications/nuclear. Significantly, the IEA notes that "this is the IEA's first report on nuclear power in nearly two decades," which underscores the degree to which nuclear power has been out of favor in the West's energy community.

6. For one example among many, see Bill Gates's talk at the Stanford Global Energy Forum, https://www.youtube.com/watch?v=d1EB1zsxWok&feature=share.

7. https://www.rusnano.com/about/press-centre/first-person/podkast-nanosvod. For the time being, Chubais's podcasts are still located on this website, despite his recent departure from Rusnano.

8. The IEA defines the "advanced economies" as consisting of the member countries of the Organisation for Economic Co-operation and Development (OECD) plus Bulgaria, Croatia, Cyprus, Malta, and Romania. Because Russia is only an associate member of the OECD, it is not included among the advanced economies under this definition. Russia is more commonly grouped with China and India as an "emerging economy."

9. IEA, *Nuclear Power in a Clean Energy System.*

10. Ed Crooks, "Nuclear Power Backers Push Cheaper, Smaller Plants," *Financial Times,* March 11, 2019, https://www.ft.com/content/b4c5ecf6-28a2-11e9-9222-7024d72222bc.

11. United States Nuclear Regulatory Commission, V. C. Summer Nuclear Station, Units 2 and 3, https://www.nrc.gov/reactors/new-reactors/col-holder/sum2.html.

12. David Keohane, "EDF Increases Hinkley Point C Nuclear Plant Costs," *Financial Times,* September 25, 2019, https://www.ft.com/content/92102452-df62-11e9-9743-db5a370481bc.

13. Geert De Clercq, "French Regulator Puts EDF Flamanville Nuclear Plant on Safety Watch," Reuters, September 11, 2019, https://www.reuters.com/article/us-edf-nuclearpower-flamanville/french-regulator-puts-edf-flamanville-nuclear-plant-on-safety-watch-idUSKCN1VW0Y5.

14. According to the International Energy Agency's (IEA) *Global Energy Outlook 2019*, annual investment in nuclear going forward is projected to be only a tiny fraction of the world's total investment in fuels and power: from $64 billion in 2019 to $74 billion per year in the 2030s (in the IEA's Stated Policies scenario) compared to $398 billion for renewables and $455 for electricity networks, and a world total in fuels and power of $1,990 billion (all in 2018 dollars). International Energy Agency, *Global Energy Outlook 2019* (Paris: 2019), p. 50, https://www.iea.org/reports/world-energy-outlook-2019.

15. IEA, *Global Energy Outlook 2019*, p. 256.

16. Ibid., p. 265.

17. Over the next two decades, in the International Energy Agency's Sustainable Development scenario, installation of small modular reactors (SMRs) could represent a market of perhaps 60 to 70 new units per year, averaging 10 megawatts each, compared to the large 1,000 MW units in use today (IEA, *Global Energy Outlook 2019*, fig. 2.25, p. 119). An interesting new development in the United States, and a possible harbinger of the future, is the recent first-ever approval for an SMR design by the Nuclear Regulatory Commission in August 2020. The project, if it goes forward, would lead to a new light-water SMR in Portland, Oregon.

18. Bloomberg New Energy Finance and McKinsey are more reserved, projecting only 9 percent for nuclear's share by 2040.

19. There are some signs, however, that the consensus in the West may be slowly evolving toward a more positive view of nuclear power as a response to climate change. See the discussion of the future of nuclear power and climate change in *Foreign Affairs*, May / June 2020.

20. See the presentation by Roman Golovin, deputy director of Rosatom's Department for Strategic Management, at the February 2020 Skolkovo School of Management's energy seminar, February 29, 2020, https://www.youtube.com/watch?v=Jg-kL_8aGhQ&list=PLZWnvmqHFxphEDUmGXmcjHp2ujmO3Hzm l&index=39. Golovin's source, as evidenced by his slides, is a blend of Western reports, ranging from Bloomberg New Energy Finance to McKinsey. The US Energy Information Administration projects 2040 global electricity demand at 33,800 terawatt-hours. The Institute of Energy Studies of the Russian Academy of Sciences, which is in some respects the godfather of the Skolkovo Energy Center, comes in at 37,500 terawatt-hours.

21. From a large literature on the Soviet period, see in particular the remarkable collection of articles, across a wide range of sectors, in Ronald Amann and Julian

Cooper, eds., *Industrial Innovation in the Soviet Union* (New Haven, CT: Yale University Press, 1982). See also Thane Gustafson, "Why Doesn't Soviet Science Do Better than It Does?," and Bruce Parrott, "The Organizational Environment of Soviet Applied Research," both in Linda Lubrano and Susan Gross Solomon, eds., *The Social Context of Soviet Science* (Boulder, CO: Westview Press, 1980), pp. 31–68 and 69–100, as well as, in the same collection, many other essays by noted scholars of Soviet science and technology.

22. Official foreign-trade statistics come from the Federal Customs Service, http://customs.ru/folder/502. According to Irina Dezhina, dependence on imports in telecommunications is 94 percent; in medical equipment, 92 percent; and in pharmaceuticals, over 90 percent. High-tech exporters themselves are dependent on import of supplies (82 percent) (personal communication).

23. V. Vlasova, L. Gokhberg, E. Dyachenko, et al., *Russian Science and Technology in Figures* (in Russian) (Moscow: National Research University, Higher School of Economics, 2018), p. 34, https://issek.hse.ru/mirror/pubs/share/215179745. The report appears to use the term "science" to include both basic and applied research.

24. Loren Graham, *Fame or Oblivion?* (forthcoming). I am grateful to Professor Graham for sharing with me a draft of this work.

25. For the argument that today's oil and gas are advanced technological enterprises, see Thane Gustafson, *Wheel of Fortune: The Battle for Oil and Power in Russia* (Cambridge, MA: Harvard University Press, 2012), chap. 13. For the story of the adoption of advanced techniques by the oil and gas industries over the last decade, combined with the continued dependence on imported technology, see Chapter 2 of this book.

26. On this key point, see the thoughtful essay by Anna Fediunina and Iuliia Aver'ianova, "Kupit', chtoby prodat'," *Ekspert,* no. 39, September 23, 2019, pp. 79–81, https://expert.ru/expert/2019/39/kupit-chtobyi-prodat/. The article makes an important distinction between forward linkages and backward linkages, the latter consisting of imports that ultimately lead to high-tech exports. In 2015 only one-quarter of Russia's exports stemmed from such backward linkages—in striking contrast to Eastern Europe, where the proportion varied between one-half and three-quarters.

27. Theodore Shabad, *Basic Industrial Resources of the USSR* (New York: Columbia University Press, 1969), p. 33.

28. For a richly detailed and deeply researched account of the early history of the Soviet civilian nuclear program and the impact of the Soviet collapse on the nuclear industry, especially its safety aspects, see Paul Josephson, *Red Atom* (Pittsburgh: University of Pittsburgh Press, 2000). I am grateful to Paul Josephson for his helpful comments on an early draft of this chapter.

29. Judith Thornton, "Soviet Electric Power in the Wake of the Chernobyl Accident," in US Congress, Joint Economic Committee, *Gorbachev's Economic Plans*, vol. 1 (Washington, DC: US Government Printing Office, 1987), p. 517.

30. US Congress, Joint Economic Committee, *Gorbachev's Economic Plans*, p. viii.

31. Igor Khripunov, "Russia's Minatom Struggles for Survival: Implications for U.S.-Russian Relations," *Security Dialogue* 31, no. 1 (March 2000): 55–69.

32. Igor Korochenko, "Otechestvennyi iadernyi kompleks razvalivaetsia," *Nezavisimaia gazeta*, April 30, 1999, accessed via East View.

33. See Iakov Pappe and Ekaterina Drankina, "Kak natsionaliziruiut Rossiiu: Atomnaia promyshlennost'," *Kommersant-Vlast'*, no. 37, September 24, 2007, p. 20, https://www.kommersant.ru/doc/806717. According to Pappe and Drankina, these private investors (who as a rule were nuclear insiders) were interested in the non-nuclear production of these factories, since there were no nuclear orders at the time. Thus, the Degtiarov Factory in Kovrov made centrifuges, but also non-nuclear products. Pappe has made a number of studies of the broad process of nationalization from 2000 onward. For an account of the complex transactions that ultimately led to the formation of Rosatom, see Iakov Pappe, "Izmenenie sootnosheniia mezhdu chastnym i gosudarstvennym sektorami v rossiiskom krupnom biznese v 2000–2013gg.: Sub'ektnyi podkhod," *Problemy prognozirovaniia*, no. 3 (2014): 32–45.

34. Alena Kornysheva, "Otchet: 'Rosatom' promakhnulsia s pribyl'iu," *Kommersant Daily*, February 17, 2005, accessed via East View.

35. For the story of Kakha Bendukidze and his attempts to take over key nuclear assets, see Pappe, "Izmenenie sootnosheniia mezhdu chastnym."

36. Pappe and Drankina, "Kak natsionaliziruiut Rossiiu."

37. Decree of the President of Russian Federation #556, 27 April 2007, "On the Restructuring of the Nuclear Energy Complex of Russia," http://kremlin.ru /events/president/news/38847.

38. Vladimir Rozhkov and E.Veselova, "'Novyi oblik' predpriiatiia atomnoi otrasli," *EKO*, no. 9 (September 2012): 126–137.

39. Ibid. Nikol'skii and Egorova, "Kirienko razdelit Rosatom."

40. International Atomic Energy Agency, "List of Member States," https:// www.iaea.org/about/governance/list-of-member-states.

41. Kirienko's meeting with Putin, September 2015, transcript available at http://kremlin.ru/events/president/news/50373.

42. Kirienko's annual meetings with Putin were regularly reported verbatim on the Kremlin website, although seemingly in shortened form.

43. Sergei Leskov, "Smert' Sredmasha," *Profil'*, November 20, 2006, accessed via East View.

44. The commission was headed by an economist, Anna Belova, who had previously overseen a similar restructuring of the railroad sector. See an interview

with Belova by Georgii Bovt, "Loshad' mozhno privesti k vode, no ee ne zastavit' pit'," *Profil'*, February 26, 2007, pp. 44–50, accessed via East View.

45. Aleksandr Kolesnichenko, "Rosatom: Kirienko privatiziruet AES?," *Argumenty i Fakty*, no.6, February 8, 2006, p. 2, accessed via East View.

46. Sergej Mahnovski and Konstantin Kovalenko, "The Revival of Nuclear Power in Russia," Cambridge Energy Research Associates Decision Brief, March 2007, p. 6 (available from the author).

47. https://rosstat.gov.ru/search?q=потребление+энергии.

48. See Russian Statistical Service (Rosstat), https://rosstat.gov.ru/search?q=потребление+энергии.

49. Indeed, by 2016, Kirienko's last year at the head of Rosatom, Russian domestic power demand was only 5.4 percent higher than it had been when he arrived eleven years earlier. "Prezident RF postavil na zasedanii Komissii po modernizatsii . . . ," July 23, 2009, https://www.atomic-energy.ru/news/2009/07/23/5093.

50. ITAR-TASS, July 22, 2014, http://atominfo.ru/newsi/p0660.htm. These numbers are reported according to Russian Accounting Standards (RAS).

51. A brief history of the Engineering Division (Inzhiningovyi Divizion) can be found on its website at https://ase-ec.ru/about/history/. It is made up of four principal units: the ASE Engineering Company (based in Nizhny Novgorod); two design institutes, Atomenergoproekt (Moscow) and Atomproekt (Saint Petersburg); and the specialized export division, Atomstroyexport (Moscow). The latter controls the construction and operation of power plants outside Russia; matters related to fuel are handled by TVEL, Rosatom's fuels division, and by Tekhsnabeksport, known outside Russia as TENEK. These supply Western-operated pressurized-water reactors with fuel services as well as Russian ones.

52. Nick Galucci, "Will the West Let Russia Dominate the Nuclear Market?," *Foreign Affairs*, August 3, 2017, https://www.foreignaffairs.com/articles/russian-federation/2017-08-03/will-west-let-russia-dominate-nuclear-market.

53. See the website of Rosatom, www.rosatom.ru.

54. Rosatom, "NPPs under Construction," https://rosatom.ru/production/design/stroyashchiesya-aes/.

55. Rosatom, "ROSATOM's Integrated Solution," https://www.rosatom.ru/en/integrated-offer/.

56. Kirienko television interview on *Russia Today*, July 3, 2012, https://www.youtube.com/watch?app=desktop&v=L9CzQSrw3yw.

57. Kirienko speaking at Medvedev's Commission, July 2009, accessed at http://kremlin.ru/events/president/trips/4887.

58. The next two paragraphs on China are based on International Atomic Energy Agency (IAEA), "China Highlights Nuclear Energy," https://www.iaea.org/newscenter/news/china-highlights-nuclear-innovation-to-meet-climate-goals

-at-iaea-conference, and on World Nuclear Association, "Nuclear Power in China," March 2021, https://www.world-nuclear.org/information-library/country-profiles /countries-a-f/china-nuclear-power.aspx.

59. Before joining Rosatom in 2012—another Kirienko legacy—Golovin worked for four years with Deloitte and two years with Booz, while earning a BA in management and an MA in finance from MGIMO. (LinkedIn, https://www .linkedin.com/in/roman-golovin-58187a5/?originalSubdomain=ru.)

60. Vera Kolerova, "'Rosatom' dobavit stoimosti," *Ekspert,* no. 50, December 10, 2018, accessed via East View.

61. Emilio Bellini, "Russian Nuclear Giant Rosatom Enters Storage Business," *PV Magazine,* October 9, 2020, https://www.pv-magazine.com/2020/10/09/russian -nuclear-giant-rosatom-enters-storage-business/. I am grateful to Philip Vorobyov for calling this item—as well as many others—to my attention.

62. See the discussion of hydrogen in Chapter 3. Rosatom's efforts in hydrogen are described in A. Mastepanov, "Vodorodnaia energetika Rossii: Sostoianie io perspektivy," *Energeticheskaia politika* (the monthly journal of the Ministry of Energy), December 23, 2020, https://energypolicy.ru/a-mastepanov-vodorodnaya -energetika-rossii-sostoyanie-i-perspektivy/energetika/2020/14/23/.

63. Kolerova, "'Rosatom' dobavit stoimosti."

64. Aleksandr Mekhanik, "Additivnye tekhnologii—eto uzhe ne fantazii," *Ekspert,* August 27, 2018, pp. 44–47, accessed via East View.

65. IPG was originally the creation of a Russian entrepreneur, Valentin Gapontsev. Gapontsev's story is similar to that of a Silicon Valley entrepreneur. He established a small business in 1990 in the basement of his institute—technically illegal at the time, but taking advantage of the increasing legal void as the Soviet Union neared collapse. He came to the attention of an Italian telecommunications company, which invited him to come to Italy. He expanded from there into Germany, and ultimately to the rest of the world. IPG has "half-returned" to Russia, in the sense that it has a Moscow office and supplies lasers to RusAT, but curiously, Rosatom hardly ever mentions its Russian origins. See Sergei Tikhonov, "Russkii additivnyi proryv," *Ekspert,* no. 12, March 20, 2017, pp. 33–39, https:// dlib-eastview-com.proxy.library.georgetown.edu/browse/doc/48469458. Loren Graham tells the story of Valentin Gapontsev in *Lonely Ideas: Can Russia Compete?* (Cambridge MA: MIT Press, 2013), pp. 88–89. IPG Photonics today is a worldwide company with headquarters in Oxford, Massachusetts. IPG maintains a Russian office under the name of IPG-IRE-Polus in Moscow Oblast (compared to five offices in China).See the IPG website at https://www.ipgphotonics.com/en /applications/materials-processing/additive-manufacturing.

66. "Pervaia v mire plavuchaia ATC razrabatyvaetsia v Peterburge," *Sankt-Petersburgskie Vedomosti,* no. 1 (2002). There was also a design for a smaller floating

nuclear power plant that would be suitable for moving into rivers and generating power there. This smaller floating vessel was supposed to be developed at the nearby "Zvezdochka" plant. Nikolai Kalistratov and Viacheslav Korb, "Perspektivy plavuchikh ATES," *Nezavisimaia gazeta,* August 14, 2007, accessed via East View.

67. For a history of the repeated delays and some of the reasons for them, see Aleksandra Gritskova, "Golovnaia PATES uplyla s 'Sevmasha,'" *Kommersant-Daily,* August 11, 2008; and Egor Popov, "Baltzavod dostroit PATES k 2016 godu," *Kommersant-Daily,* August 24, 2012, both accessed via East View.

68. The ship generates power and heat sufficient for 100,000 people. See the Rosatom website, May 22, 2020, https://rosatom.ru/journalist/news/edinstvennaya -v-mire-plavuchaya-atomnaya-teploelektrostantsiya-vvedena-v-promyshlennuyu -ekspluatatsi/?sphrase_id=1385807. See also Irina Gagarinskaia and Oleg Nikoforov, "Plavuchie energobloki meniaiut kartinu energosnabzheniia strany," *Nezavisimaia gazeta,* February 11, 2020, accessed via East View.

69. CNNC is the Chinese National Nuclear Corporation. In 2016 CNNC established a subsidiary, the China Rich Energy Corporation, to develop wind and power.

70. Anastasiia Fomicheva and Natal'ia Skorlygina, "PATES doplyvut do Kitaia," *Kommersant,* July 30, 2014, accessed via East View.

71. Sofia Samokhina et al., "Iadernye dostizheniia priniali geroicheskuiu formu," *Kommersant,* July 5, 2018, p. 3, https://www.kommersant.ru/doc/3676874. Evidently there were three reasons for the award. The first was Rosatom's contribution to a new generation of military applications—Rosatom is still a major player in Russia's military-industrial system. Second was Kirienko's effective performance as the manager of Putin's 2018 presidential campaign. But the third major reason for the award, according to Samokhina, was Kirienko's transformation of Rosatom into a major exporter of Russian civilian nuclear technology. Several members of Kirienko's team were also awarded decorations at the same time.

72. Christopher De Vere Walker, "Russia's Growing Nuclear Power Prowess Home and Away: Setting the Agenda for the Next Generation Nuclear Reactors," *IHS Markit Strategic Report,* February 26, 2019.

73. Jane Nakano, "The First-Ever U.S. Approval for Small Modular Reactor Design and Its Implications," Center for Strategic and International Studies, September 17, 2020, https://www.csis.org/analysis/first-ever-us-approval-small -modular-reactor-design-and-its-implications?utm_source=CSIS+All&utm _campaign=384648f7b2-EMAIL_CAMPAIGN_2018_08_31_06_36_COPY _01&utm_medium=email&utm_term=0_f326fc46b6-384648f7b2-222100893.

74. SMRs have their critics, however, who point to a number of downsides and doubt that SMRs can play a significant role in fighting climate change. For a detailed critique of SMRs, see Arjun Makhijani and M. V. Ramana, "Why Small

Modular Nuclear Reactors Won't Help Counter the Climate Crisis," Environmental Working Group (EWG), March 2021, https://www.ewg.org/news-insights/news /why-small-modular-nuclear-reactors-wont-help-counter-climate-crisis.

7. Russia's Agricultural Renaissance

1. Several poignant books have been written about the interaction between Soviet industrialization and collectivization. Among others, two meticulous and systematic works present the devastating aftermath: Robert Conquest, *Harvest of Sorrow: Soviet Collectivization and the Terror-Famine* (Oxford: Oxford University Press, 1987), and Anne Applebaum, *Red Famine: Stalin's War on Ukraine* (New York: Anchor Books, 2018).

2. Throughout this chapter, areas are given in hectares. One hectare is a 1/100 of a square kilometer, or 2.47 acres.

3. A. I. Bedritskii, ed., *Natsional'nyi doklad: Global'nyi klimat i pochvennyi pokrov Rossii,* vol. 1 (Moscow: Pochvennyi Institut im. V.V. Dokuchaeva, 2018), pp. 72–94, http://www.esoil.ru/publications/books/news26032018.html. It was only after 2007 that the sown area began to recover somewhat. The latest figure is 80 million hectares in 2019.

4. Despite the publicity given to the Russian "countersanctions," they actually apply only to some Western countries, not the world. The total number is just over 30 nations. The other 160+ countries are open for business, and, in fact, Russia has increased food trade with Asia (especially China), the Middle East, and South America. African nations are also being courted, as are nations in South Asia (India, Pakistan, Bangladesh). Food imports from China are up substantially. Food trade within the Eurasian Economic Union has also increased. So there has been substitution for Western food. But the most important consequence of the countersanctions has been the stimulation of domestic production, as described below. (I am indebted for this point to Professor Stephen Wegren, in a private communication.)

5. See the discussion of the outlook for grain production and exports in Stephen K. Wegren, "Can Russia's Food Exports Reach $45 Billion in 2024?," *Post-Communist Economies* 32, no. 2 (2020): 147–175, esp. the sections "Grain: The Positive" and "Grain: The Negative," pp. 157–159, https://doi.org/10.1080/14631377.2019.1678346. However, the discussion of the negatives focuses mainly on near-term downsides. For Wegren's observations on the possible longer-range impacts of climate change, see Stephen K. Wegren with Alexander Nikulin and Irina Trotsuk, *Russia's Food Revolution: The Transformation of the Food System* (New York: Routledge, 2021), chap. 5, pp. 42ff.

6. United Nations, Food and Agriculture Organization (FAO), *Agriculture and Climate Change: Challenges and Opportunities at the Global and Local Level* (Rome: FAO, 2019), p. 3, http://www.fao.org/3/CA3204EN/ca3204en.pdf.

7. EPA, *Global Greenhouse Gas Emissions Data,* https://www.epa.gov/ghg emissions/global-greenhouse-gas-emissions-data#Sector.

8. Ivan Buzdalov, "Sovremennoe polozhenie v sel'skom khoziaistve Rossii: Sistemnyi agrarnyi krizis prodolzhaetsia," *Obshchestvo i ekonomika,* no. 3 (2018): 78, http://naukarus.com/sovremennoe-polozhenie-v-rossiyskom-selskom-hozyaystve -i-novye-trebovaniya-k-agrarnoy-politike. Buzdalov is a senior researcher at the Institute of Economics of the Russian Academy of Sciences. He does not cite a source for these numbers.

9. There is an important distinction to be made between "sustainable agriculture" and "organic farming." True sustainable agriculture involves preserving the soil or maintaining fertility by using updated practices for plowing or cropping. Organic farming, on the other hand, merely involves the production of food without the use of chemical fertilizers, antibiotics, and pesticides. The Russian government has announced ambitious targets for organic farming, projecting that it can capture 10 to 15 percent of the organic food trade by 2030.

10. Tat'iana Kulistikova, "Pogoda stanovitsia nervnoi," *Agroinvestor,* September 4, 2019, pp. 1–13, https://www.agroinvestor.ru/analytics/article/32343-pogoda-sta novitsya-nervnoy/.

11. Quoted in Kulistikova, "Pogoda stanovitsia nervnoi."

12. Kulistikova, "Pogoda stanovitsia nervnoi."

13. The worst drought in recent years came in 2010, with lesser droughts in 2012, 2018, and 2019.

14. Stephen K. Wegren and Alexander M. Nikulin, "Food and Foreign Policy," in Stephen K. Wegren, ed., *Putin's Russia: Past Imperfect, Future Uncertain,* 7th ed. (Lanham, MD: Rowman and Littlefield, 2019), p. 271.

15. Putin presentation at meeting of the Commission on Military-Technical Co-operation, https://tass.com/defense/992835, March 2018.

16. Wegren and Nikulin, "Food and Foreign Policy," pp. 269–290. In 2019, agriculture minister Dmitry Patrushev reported proudly that agricultural exports had reached \$25.9 billion in 2018, twice the level of arms exports (https://regnum .ru/news/economy/2590896.html). The original source is Patrushev's online column on "Pravitel'stvo onlain," at ria.ru.

17. More recently, the pandemic has caused the Russian leadership to reevaluate the relative priority of exports vs. food security. The Russian government has introduced new export quotas and tariffs on grain and oilseeds, effective in February 2021, with the aim of limiting exports, and in his annual press conference in December 2020, Putin said it had been a mistake to subsidize food exports. In parallel, the government has imposed price restraints on key domestic foods such as vegetable oils and sugar. So long as the pandemic lasts, in short, concerns over domestic food security prevail over export ambitions. (I am grateful to Professor Wegren for calling my attention to these developments.)

18. The countersanctions ran out at the end of 2020, but they are likely to be extended, in view of their importance for domestic production.

19. Rosgidromet, *Vtoroi otsenochnyi doklad Rosgidrometa ob izmeneniiakh klimata i ikh posledstviiakh na territorii Rossiiskoi Federatsii,* http://downloads.igce.ru/publications/OD_2_2014/v2014/htm/.

20. Bedritskii, *Natsional'nyi doklad,* 1:47.

21. See note 3 above, and also references to Bedritsky's role in Chapter 1.

22. Bedritskii, *Natsional'nyi doklad,* 1:48.

23. Commentary by Academician Andrei Paptsov, "Strategiiu razmeshcheniia zernovykh i kormovykh kul'tur izmenit klimat," *Krest'ianskie Vedomosti,* November 21, 2018, https://kvedomosti.ru/news/kommentarij-akademik-andrej-papcov-strategiyu-razmeshheniya-zernovyx-i-kormovyx-kultur-izmenit-klimat.html/print/. The website of the Institute describes its massive atlas of climate zones and changing natural conditions, together with their impact of yields, compiled as of 2018. See http://www.cxm.obninsk.ru/index.php?id=77.

24. A good summary of the range of Russian views on the possible positive and negative impacts of climate change on the Russian economy, including agriculture, is Vladislav Grinkevich, "Smozhet li Rossiia zarabotat' na global'nom poteplenie?," *Profil',* no. 5 (February 2020): 32–37, https://dlib.eastview.com/browse/doc/57336258. On the whole, Grinkevich tends to give the negatives greater weight.

25. The point is repeated several times throughout the report. This edition was published under the overall editorship of R. Edel'geriev, chief advisor to President Putin on environment and climate change, and is the result of an extensive inventory of the country's soils.

26. Quoted in Elena Bukovskaia, "Gidromettsentr: Klimat ne pozvolit uvelichit' ploshchad' sel'khozzemel'," RosNG.ru, February 27, 2020, https://www.rosng.ru/post/klimat-ne-pozvolit-uvelichit-ploshad-selhozzemel-gidrometcentr?fbclid=IwAR3ih9J9Vnd4pYerWuBHlxVoa692x8U47OZ5YOLvEcmPzyAiIboXFmPJapk.

27. Quoted in Vera Kolerova, "'Rosatom' dobavit stoimosti," *Ekspert,* no. 50, December, 10 2018, accessed via East View.

28. Volume 1 of Bedritskii, *Natsional'nyi doklad,* was followed in 2019 by volume 2, under the same title but under the overall editorship of Ruslan Edel'geriev (http://www.esoil.ru/publications/books/nacdoklclimat.html). The reference here is to the 2019 volume, pp. 385ff.

29. The five top grain-producing oblasts are (in order): Krasnodar, Rostov, Stavropol, Voronezh, and Altai, according to Agroinvestor, https://www.agroinvestor.ru/, various issues.

30. Edel'geriev, *Natsional'nyi doklad,* 2:385. This fact helps put in perspective the fact that grain production did not regain Soviet levels until 2016: by that time it was being conducted on a much smaller land area, but with a much higher level of productivity. (See Edel'geriev, *Natsional'nyi doklad,* fig. 5.13.) For discussion of

the changes in land use since the Soviet period, see T. F. Nefedova, "Twenty-Five Years of Russia's Post-Soviet Agriculture: Geographical Trends and Contradictions," *Regional Research of Russia* 7, no. 4 (2017): 311–321, https://ideas.repec.org /a/spr/rrorus/v7y2017i4d10.1134_s2079970517040074.html. I am grateful to Dr. Nefedova for making this article available to me.

31. Figures from Russian foreign trade statistics. See http://www.ved.gov.ru /monitoring/foreign_trade_statistics/basic_goods_export/. Updated figures for 2020 found at http://customs.gov.ru/press/federal/document/246946. In January– June 2020, Russia exported roughly $5.6 billion worth of food, and imported around $7.1 billion worth.

32. Professor Wegren has written many important books and articles on Russian agriculture over the past two decades, and he is rightly considered the West's foremost expert on the subject. On the origins of the present system of landholding, see Stephen K. Wegren, *Land Reform in Russia: Institutional Design and Behavioral Responses* (New Haven, CT: Yale University Press, 2009), followed at about the same time by Wegren, "Agriculture in the Late Putin Period and Beyond," in Stephen K. Wegren and Dale R. Herspring, eds., *After Putin's Russia: Past Imperfect, Future Uncertain,* 4th ed. (Lanham MD: Rowman and Littlefield, 2010), pp. 199–222.

33. The full study, by V. Uzun, N. Shagaida, and Z. Lerman, is not yet publicly available. Its findings are summarized in a forthcoming article, "Russian Agroholdings and Their Role in Agriculture." I am grateful to Professor Wegren for making a draft available to me.

34. Quoted in Wegren, "Agriculture in the Late Putin Period," p. 210.

35. Shagaida, as cited in Wegren, *Land Reform in Russia,* chap. 5, n10.

36. V. Uzun, N. Shagaida, and Z. Lerman, "Russian Agroholdings."

37. Stephen K. Wegren, *Food Policy and Food Security: Putting Food on the Russian Table* (Lanham MD: Lexington Books, 2018), pp. 17–18.

38. Uzun, Shagaida, and Lerman, "Russian Agroholdings."

39. Tat'iana Kulistikova, "Verbal'nye investitsii prodolzhaiutsia," *Agroinvestor,* March 1, 2019, https://www.agroinvestor.ru/investments/article/31306-verbalnye -investitsii-prodolzhayutsya/.

40. According to a summary by the US Department of Agriculture, "174,800 registered 'peasant' farms (averaging about 600 acres apiece, or 243 hectares) account for 13 percent of total production. Nonexistent in Russia in 1990, this category of agricultural producers (*krest'ianskie,* in Russian) specializes in the production of grain, oilseeds, and other industrial crops that require high levels of mechanization." See USDA, "Russian Federation: Agricultural Economy and Policy Report" (Moscow, July 2018), https://apps.fas.usda.gov/newgainapi/api/report /downloadreportbyfilename?filename=Agricultural%20Economy%20and%20 Policy%20Report_Moscow_Russian%20Federation_7-19-2018.pdf.

41. Anatolii Kostyrev, "Inostrantsy prosypali zerno," *Kommersant,* February 28, 2020, accessed via Factiva (https://professional.dowjones.com/factiva/). The share of foreign traders was more than one-half in 2013, but by 2020 it had declined to less than one-third.

42. The sudden rise of Steppe as a grain exporter has been followed with fascination by the Russian media. See in particular the reporting of Anatolii Kostyrev in *Kommersant*—for example, "'Step' vyshla k reke," *Kommersant,* November 27, 2018, and "V 'Stepi' pokazalsia eksport," *Kommersant,* August 27, 2018, both accessed via Factiva. Other agroholdings are going down the same path, such as Agronova, whose parent company is Region, which reportedly is connected to Rosneft. See Aleksei Polukhin and Anatolii Kostyrev, "Zerno doroslo do treidinga," *Kommersant,* April 23, 2019, accessed via Factiva. At last report RIF has maintained its lead as Russia's top grain exporter. See Inna Ganenko, "Top-20 eksportery postaviat za rubezh bolee 76% rossiiskogo zerna," *Agroinvestor,* June 25, 2020, https://www.agroinvestor.ru/analytics/news/33917-top-20-eksporterov-postavyat-za-rubezh-bolee-76-rossiyskogo-zerna/.

43. Anatolii Onegov, "Kak vosstanovit' plodorodie pochvy," *Nauka i zhizn',* September 2020, https://www.nkj.ru/archive/articles/6481/.

44. Kulistikova, "Verbal'nye investitsii prodolzhaiutsia."

45. Except in rhetoric. On their websites the agroholdings play up their commitment to sustainability, going green, being environmentally conscious, etc. Whether it is true or not is a different question, but from a PR standpoint they are at least aware of the growing visibility of these issues.

46. The potential benefits of information technologies are clearly on the minds of the new owners, although it is difficult to tell how much real action is taking place. See Il'ia Dashkovskii, "High-tech na zemliu: Pochemu sel'skoe khoziaistvo prodolzhaet ostavat'sia nezavisimym ot tsifrovizatsiia," *Agroinvestor,* July 8, 2020, https://www.agroinvestor.ru/analytics/article/33998-high-tech-na-zemlyu-pochemu-selskoe-khozyaystvo-prodolzhaet-ostavatsya-nezavisimym-ot-tsifrovizatsii/.

47. Khvorostina interview, *Agroinvestor,* July 17, 2020.

48. Georgy Safonov is director of the Center for the Economics of the Environment and Natural Resources of the Higher School of Economics; see https://www.hse.ru/org/persons/512771. See the interview with Safonov in Elena Kudriavtseva, "V zone riskovannogo potepleniia," *Kommersant,* March 10, 2020, https://www.kommersant.ru/doc/4259400.

49. One of the goals Putin explicitly declared in his famous manifesto "Russia at the Turn of the Millennium" is "modern agrarian policy." His exact statement is: "The revival of Russia will be impossible without the revival of the countryside and agriculture." See the full text of the manifesto at https://pages.uoregon.edu/kimball/Putin.htm.

50. For a description of the first decade of the Putin program, see Stephen K. Wegren, "Agriculture in the Late Putin Period and Beyond," in Wegren and Herspring, *After Putin's Russia*, pp. 199–222.

51. Wegren, *Food Policy*, p. xxix. In 2018, total federal spending on agriculture, as given in the annual report of the Ministry of Agriculture, was 242 billion rubles (*2018 Annual Report*, p. 12), down slightly from 248 billion in 2017 (*2017 Annual Report*, p. 6), but up from 223 billion in 2016 (*2016 Annual Report*, p. 6) and 234 billion in 2015 (*2015 Annual Report*, p. 11). The impact of unfavorable geopolitical and economic events in 2014 was evidently severe; total agricultural spending was only 189 billion rubles in that year (*2014 Annual Report*, p. 9). As for the planned growth in federal spending on agriculture in 2019–2020 and in coming years, see E. Razumnyi, "V 2020 na gosprogrammu APK predusmotreno 290 mrd. rublei," *Agroinvestor*, September 26, 2019, https://www.agroinvestor.ru /markets/news/32471-na-gosprogrammu-apk-predusmotreno-290-mlrd/.

52. Wegren, *Food Policy*, p. xiv and chap. 1.

53. V. V. Rau and E. Iu. Frolova, "Agrarnyi sector: Novye tochki rosta," *Problemy prognozirovaniia*, no. 2 (2019): 63–72.

54. "Otgruzki otechestvennoi sel'khoztekhniki na rossiiskii rynok vyrosli v 2019 na 1%." *Sel'khoiziastvennye vesti*, February 11, 2020, https://agri-news.ru /novosti/otgruzki-otechestvennoj-selxoztexniki-na-rossijskij-ryinok-vyirosli-v -2019-godu-na-1.html. In May 2020 Putin announced that subsidies for the purchase of agricultural machinery would be increased by one-third over 2019, to 18.5 billion rubles (or roughly $260 million). In parallel, there is also a smaller leasing program for agricultural machinery. "Putin podderzhal uvelichenie v 2020 godu subsidii na zakupku rossiiskoi sel'khoztekhniki," *Praim*, May 20, 2020, https:// 1prime.ru/state_regulation/20200520/831481635.html.

55. Wegren, *Food Policy*, p. 198.

56. Uzun, Shagaida, and Lerman, "Russian Agriculture," p. 11.

57. Nadezhda Kanygina, "Asimmetriia gospodderzhki," *Agroinvestor*, July 21, 2020, https://www.agroinvestor.ru/column/nadezhda-kanygina/34065-asimmetriya -gospodderzhki/.

58. Uzun, Shagaida, and Lerman, "Russian Agriculture," p. 12.

59. Quoted in Elena Maksimova and Tat'iana Kulistikova, "Agrosektoru ne khvataet finansirovaniia," *Agroinvestor*, July 11, 2020, https://www.agroinvestor.ru /analytics/news/34014-agrosektoru-ne-khvataet-finansirovaniya/.

60. See Thane Gustafson, *Reform in Soviet Politics: Lessons of Recent Policies on Land and Water* (Cambridge: Cambridge University Press, 1981), esp. chaps. 2 and 9. For the best overview of Soviet-era agriculture, see Stephen K. Wegren, *Agriculture and the State in Soviet and Post-Soviet Russia* (Pittsburgh: University of Pittsburgh Press, 1998).

61. Mikhail Privezentsev, acting director of the National Council of Grain Producers, cited in Kulistikova, "Verbal'nye investitsii prodolzhaiutsia."

62. Bedritskii, *Natsional'nyi doklad,* 1:366, table 5.27.

63. Ministry of Agriculture, "Itogi realizatsii (2014–17) federal'noi tselevoi programmy 'Razvitie melioratsii zemel' sel'skokhoziaistvennogo nazhnacheniia Rossii na 2014–2020 gody," Moscow, 2018, http://mcx.ru/upload/iblock/61d/61d4 30039b8863186a4fbb1f6ofab1c6.pdf.

64. "Itogi realizatsii," p. 20. The numbers are difficult to interpret, however. Another table, two pages later, suggests that state support for reclamation averages only about 10 billion rubles per year, with only 20 percent coming from the federal budget (p. 21). Still another series of numbers appears in the annual reports of the Ministry of Agriculture. The total there is roughly in the average range of about 20 billion rubles per year (although with wide variations from year to year), but the share of the federal budget is only about 35 percent. See "Natsional'nyi doklad o khode i rezul'tatakh v 2018 godu gosudarstvennoi programmy razvitiia sel'skogo khoziaistva," p. 191, table 9.1.1, http://mcx.ru/upload/iblock/61d/61d430 039b8863186a4fbb1f6ofab1c6.pdf.

65. "Itogi realizatsii," p. 104.

66. Ibid., p. 109.

67. Maksimova and Kulistikova, "Agrosektoru."

68. On the strengths and weaknesses of the agricultural system, which give important indications of its likely resilience in coming years, see Stephen K. Wegren, Alexander M. Nikulin, and Irina Trotsuk, "Russian Agriculture during Putin's Fourth Term: A SWOT Analysis," *Post-Communist Economies* 31, no. 4 (2019): 419–450.

69. Professor Wegren argues these two points strongly in the final chapter of *Russia's Food Revolution* (pp. 203–210) in which he addresses the coming challenges of climate change for the food system.

70. Wegren, "Food and Foreign Policy," pp. 269–290. One question for the future is how Russia will choose if it is forced to decide between domestic food security and geopolitical influence from food aid overseas. For recent developments at the end of 2020, see note 17.

71. For the outlook on the eve of the pandemic, see Stephen K. Wegren, "Can Russia's Food Exports Reach $45 Billion in 2024?," *Post-Communist Economies* 32, no. 2 (2020): 147–175.

8. A Tale of Two Arctics

1. D. P. Koptev, "Noril'skii razliv: Uroki i posledstviia," *Burenie i Neft',* July–August 2020, https://burneft.ru/archive/issues/2020-07/3.

2. "Stroitel' 'Nornikelia': 'Bol'shinstvo sovetskikh sooruzhenii na Severe— pod ugrozoi,'" *Novye izvestiia* (web version), accessed via East View (https://dlib .eastview.com/).

3. "Permafrost" is officially defined as ground that remains frozen for more than two years. However, even in the coldest seasons there is an "active layer" that melts and freezes with the seasons. This active layer constitutes what is called melting permafrost as it progressively reaches deeper with warmer winters.

4. For their comments on earlier drafts of this chapter, I am grateful to several colleagues and friends who are knowledgeable on the Russian Arctic, notably Scott Stephenson, Robert Orttung, Arild Moe, Valery Kryukov, Harley Balzer, and Marjorie Mandelstam Balzer.

5. See Rosgidromet, "Doklad ob osobennostiiakh klimata na territorii RF za 2019 god," Moscow, 2020, https://meteoinfo.ru/images/news/2020/03/12/0-klimate-rf -2019.pdf.

6. See Dmitry Streletskiy and Nikolay Shiklomanov, "Russian Arctic Cities through the Prism of Permafrost," in Robert Orttung, ed., *Sustaining Russia's Arctic Cities: Resource Politics, Migration, and Climate Change* (New York: Berghahn Books, 2017), pp. 201–220. There is also an opposite migration toward the growing oil and gas cities along the Arctic coast, but that arises primarily from sources outside the Arctic.

7. The classic Western work on the geography of industrial materials of the Soviet Union is Theodore Shabad, *Basic Industrial Resources of the USSR* (New York: Columbia University Press, 1969). This work was further developed by regular research publications by a group of American geographers led by Theodore Shabad, Matthew J. Sagers, and others. See, for example, Theodore Shabad and Matthew J. Sagers, *The Chemical Industry in the USSR: An Economic Geography* (Boulder CO: Westview Press, 1990), as well as the journal *Soviet Geography.*

8. Boris Revich, ed., *Climate Change Impact on Public Health in the Russian Arctic* (Moscow: United Nations in Russia, January 2008), https://www.researchgate .net/publication/281931888_Climate_Change_Impact_on_Public_Health_in_the _Russian_Arctic.

9. Streletskiy and Shiklomanov, "Russian Arctic Cities."

10. There was a massive outmigration in the 1990s, which has since slowed. There are major differences in migration trends among the various northern cities. Colin Reisser distinguishes three types: energy cities; cities with more diversified economies, which are growing; and older mining centers, which are still tending to decline. Thus, there are, in effect, "two Arctics." The overall result has been a decline in the working-age population, which makes the region increasingly

dependent on immigration, mostly from the non-Russian republics of the former Soviet Union. See Colin Reisser, "Russia's Arctic Cities: Recent Evolution and Drivers of Change," in Orttung, *Sustaining Russia's Arctic Cities,* chap. 1. Trends in emigration and immigration in the Russian Arctic are notoriously difficult to study, but see the excellent essays: Timothy Heleniak, "Boom and Bust: Population Change in Russia's Arctic Cities," and Marlene Laruelle, "Assessing Social Sustainability: Immigration to Russia's Arctic Cities," both in Orttung, *Sustaining Russia's Arctic Cities,* chaps. 4 and 5.

11. "V Sibiri taet merzlota," *Pravda,* June 28, 2019, p. 7, accessed via East View.

12. "V ozhidanii potopa: Chem grozit taianie vechnoi merzloty na russkom severe," *Novye izvestiia* (web version), April 29, 2020, accessed via East View. Data from the Laboratory for the Geo-Ecology of the North of the Geography Faculty of Moscow State University, which is currently conducting a major study of coastal erosion.

13. Interview with Aisen Nikolaev, the governor of Sakha, *Vedomosti,* May 29, 2019, p. 8, accessed via East View. The problem of flooding is aggravated by the fact that many of the northern towns are built in river valleys. See Oleg Anisimov and Vasily Kokorev, "Cities of the Russian North in the Context of Climate Change," in Orttung, *Sustaining Russia's Arctic Cities,* pp. 141ff.

14. Technically, Yakutsk is located to the south of the zone traditionally considered to be the Arctic. But in its climate and permafrost conditions, it fully qualifies.

15. Svetlana Sukneva and Marlene Laruelle, "A Booming City in the Far North: Demographic and Migration Dynamics of Yakutsk, Russia," *Sibirica: The Journal of Siberian Studies* 18, no. 3 (Winter 2019): 9–28, https://www.academia.edu /43393071/_A_Booming_City_in_the_Far_North_Demographic_and_Migration _Dynamics_of_Yakutsk_Russia_Sibirica_The_Journal_of_Siberian_Studies_no _3_9_28_co_authored_with_Svetlana_Sukneva_2019.

16. Ibid., p. 9.

17. A valuable source of information and links on the sustainability of the Russian Arctic cities is the website of the Promoting Urban Sustainability in the Arctic program at George Washington University, led by Professors Robert Orttung and Marlene Laruelle: https://blogs.gwu.edu/arcticpire/.

18. The leading Western authority on the Indigenous Peoples of the Russian Far North and East (especially Sakha) is Professor Marjorie Mandelstam Balzer of Georgetown University, where she is co-coordinator of the Indigenous Studies working group (https://indigeneity.georgetown.edu). Her latest work is an edited collection, *Arctic Issues and Identities,* issue no. 4 of *Anthropology and Archeology of Eurasia* 58 (2019), with an introductory essay by Professor Mandelstam Balzer on pp. 1–9. Her next book is *Galvanizing Nostalgia? Indigeneity and Sovereignty in Siberia* (Ithaca, NY: Cornell University Press, 2022). I am indebted to Professor

Mandelstam Balzer and to Professor Harley Balzer for calling my attention to many valuable sources on the region.

19. On the consequences of climate change for the energy infrastructure of the Arctic, see Liudmila Nefedova, "Adaptatsiia energokompleka k izmeneniiam klimata v Arktike," *Energeticheskaia politika* (a monthly publication of the Russian Ministry of Energy), no. 9 (September 2020), https://energypolicy.ru/l-nefedova-adaptacziya-energokompleks/energetika/2020/16/10/.

20. On the rising political activism of people in northern oblasts and krais, particularly in Sakha, see the work of Professor Marjorie Mandelstam Balzer of Georgetown University. See especially Marjorie Mandelstam Balzer, "Indigeneity, Land and Activism in Siberia," in A. Tidwell, ed., *Land, Indigenous Peoples and Conflict* (New York: Routledge, 2017), pp. 9–27; and also Vera Solovyeva, "Ecology Activism in the Sakha Republic: Russia's 'Large Numbered' Indigenous Peoples and the United Nations Declaration of the Rights of Indigenous Peoples," in Pamela Cala and Elsa Stamapatalou, eds., *Walking and Learning with Indigenous Peoples* (New York: Columbia University Press, 2018), pp. 119–139. See also Arbakhan K. Magomedov, "How Native Peoples of Russia's Arctic Defend Their Interests," *Anthropology and Archeology of Eurasia* 58, no. 4 (2019): 215–245, https://www.tandfonline.com/doi/abs/10.1080/10611959.2020.1811560.

21. Oleg Anisimov and Robert Orttung, "Climate Change in Northern Russia through the Prism of Public Perception," *Ambio* 47, no. 6 (October 2018), https://doi.org/10.1007/s13280-018-1096-x. For a comprehensive examination of Russian policy responses at both the federal and regional levels, see Orttung, *Sustaining Russia's Cities.*

22. Norilsk was the single most important cargo producer for the NSR in Soviet times. The Soviet nuclear icebreaker fleet was developed largely to escort traffic to and from Dudinka, the loading port for Norilsk, at the mouth of the Yenisey River. From 1978 this was done on a year-round basis, with nonferrous metals being transported via the Kara Sea to Murmansk, and sometimes directly to European destinations. (I am grateful to Arild Moe of the Nansen Institute for these points.)

23. See the outspoken criticism from the noted Siberian economist, academician Valeriy Kriukov, head of the Novosibirsk branch of the Russian Academy of Sciences, in a joint interview with geologist N. P. Pokhilenko, "Kak defragmentirovat' Arktiku," *Nauka v Sibiri,* January 2020, http://www.sbras.info/articles/opinion/kak-defragmentirovat-arktiku-vzglyad-ekonomista-i-geologa.

24. The Northern Sea Route has long excited the imagination of geographers and explorers, and there is a large literature about it. See, for example, Marcus Matthias Keupp, ed., *The Northern Sea Route: A Comprehensive Analysis* (Wiesbaden: Springer Verlag, 2015).

25. Technically the Northern Sea Route does not include the Barents Sea, but there have been proposals to expand the official designation of the route to include the stretch between Novaya Zemlya and Murmansk. Putin himself has long defined the NSR as extending from Saint Petersburg and Murmansk to Chukotka. (See his speech of April 2000 in Murmansk, http://kremlin.ru/events/president/transcripts/21346.) But at this writing these proposals have not been adopted.

26. See "Ice-Breaking LNG Carrier Makes First Eastwards Voyage from Yamal LNG," Lloyd's List, https://lloydslist.maritimeintelligence.informa.com/LL1123326/Ice-breaking-LNG-carrier-makes-first-eastwards-voyage-from-Yamal-LNG.

27. For background on the history of the Northern Sea Route and developments up to 2010, see Arild Moe, "The Northern Sea Route: Smooth Sailing Ahead?," *Strategic Analysis* 38, no. 6 (2014): 784–802, https://www.tandfonline.com/doi/abs/10.1080/09700161.2014.952940.

28. On the potential benefits and liabilities for Russia from climate change, see Vladislav Grinkevich, "Smozhet li Rossiia zarabotat' na global'nom poteplenii?," *Profil'*, no. 5, February 10, 2010, pp. 32–37.

29. Address to the State Council and the Maritime Kollegium, May 2, 2007, http://kremlin.ru/events/president/transcripts/24224. He was apparently unconscious of the irony that he was implicitly acknowledging the reality of climate change.

30. Arctic Forum 2019, http://kremlin.ru/events/president/news/60250.

31. The impact of the COVID-19 pandemic is as yet uncertain. Recent commentary suggests that the entire project is being scaled back. See "Rukhnul ocherednoi strategicheskii megaproekt," *Sovetskaia Rossiia,* September 19, 2020, http://www.sovross.ru/articles/2026/49797?mc_cid=64d2becc6f&mc_eid=5807b6d712.

32. Press conference by Osama Rabie, chairman of the Canal Authority, Reuters staff report, August 6, 2020, https://www.reuters.com/article/egypt-econ omy-suezcanal/egypts-suez-canal-revenues-up-4-7-in-last-5-years-chairman -idUSL8N2F84GW.

33. On the evolution of Russia's Arctic policy, see an analysis by one of the West's leading experts, Arild Moe of the Nansen Institute in Oslo, "A New Russian Policy for the Northern Sea Route?," *The Polar Journal* 10, no. 2 (2020): 209–227, https://doi.org/10.1080/2154896X.2020.1799611.

34. Arild Moe has focused his recent research on establishing the costs of the Northern Sea Route, and especially shipping. See "Ekonomika Arktiki: Chego ne khvataet? Primer Severnogo Morskogo Puti," *EKO,* no. 12 (2020): 62–83. Moe is particularly critical of the failure of the various government agencies involved to conduct a systematic cost-benefit analysis of the NSR or to release the details of their plans for public discussion.

35. Statement by Aleksandr Krutikov, deputy minister for the North and Far East, at the Ninth International Forum, "Arctic: Today and the Future," *PortNews*, April 3, 2020, https://en.portnews.ru/news/293938/. In 2020 the Duma's Chamber of Accounts (*Schetnaia Palata*), the lower house's watchdog over the government's programs, issued a report in which it cast doubt that the state's goal of 80 MMt / y of traffic via the NSR by 2024 would be achieved. The principal author of the report, V. N. Bogomolov, is a specialist on transportation issues who monitors the implementation of transportation plans by the government ministries. See his brief biography and a list of his reports at https://ach.gov.ru/structure/bogomolov -valeriy-nikolaevich. "Kratkie itogi ekspertno-analiticheskogo meropriiatiia, "Monitoring khoda realizatsii meropriiatiia kompleksnogo plana modernizatsiia i rasshireniia magistral'noi infrastruktury na period do 2024 goda," https://ach .gov.ru/upload/iblock/772/77228f831f35f05e7d7f4f428665d40f.pdf. See also Moe, "Ekonomika Arktiki."

36. One much-publicized example is the delays in completing the first Russian aircraft carrier, the *Admiral Kuznetsov*, which have pushed back the nuclear icebreaker program. See *Novaia Gazeta*, December 2019, https://novayagazeta.ru /articles/2019/12/14/83167-boets-arkticheskogo-fronta. For updates on various problems and delays, see the regular reports by Paul Goble, who writes for the Jamestown Foundation's *Eurasia Daily Monitor* (*EDM*). See, for example, "Delays, Disasters and Cost Overruns Plague Putin's Projects in Arctic," *EDM*, June 4, 2020, https://jamestown.org/program/delays-disasters-and-cost-overruns-plague -putins-projects-in-arctic/.

37. Putin's remarks are available on the Kremlin website at http://kremlin.ru /events/president/news/61578. Putin did not name the shipyards that were involved. He acknowledged that military industry had long been in bad financial shape and was deeply in debt, and that this is one of the main reasons that it had not done a better job on civilian production.

38. In July 2020 the CEO of Rosatom, Aleksei Likhachev, reported to President Putin that the first of the new series, named *Arktika*, would be commissioned in October 2020, http://kremlin.ru/events/president/news/63517.

39. Marina Nabatnikova, "U atomnogo flota poiavitsia 'Lider,'" *Argumenty i Fakty*, no. 18, April 29, 2020, p. 7, accessed via East View.

40. A much-noted article by Abrahm Lustgarten of ProPublica, which appeared in the *New York Times Magazine* in December 2020, argues that as the permafrost melts, the Russian Arctic will turn into a major agricultural producer and exporter. This cannot be ruled out, but it is clearly a scenario for the more distant future. See "How Russia Wins the Climate Crisis," *New York Times Magazine*, December 21, 2020, https://www.nytimes.com/interactive/2020/12/16/maga zine/russia-climate-migration-crisis.html.

41. For a thoughtful recent analysis of the politics of the Northern Sea Route as of 2020, see Arild Moe, "A New Russian Policy."

9. Metals

1. For excellent background on Rusal and the aluminum industry, see Stephen Fortescue, "The Russian Aluminum Industry in Transition," *Eurasian Geography and Economics* 47, no. 1 (2006): 76–94; see also Ellie Martus, "Russian Industry Responses to Climate Change," *Climate Policy* 19, no. 1 (2019): 17–29. Aluminum in the Soviet era is treated in Theodore Shabad, *Basic Industrial Resources of the USSR* (New York: Columbia University Press, 1969), pp. 58–65.

2. The only significant exception is Alrosa, the world's largest diamond producer.

3. The roles of China and India have been especially significant since 2000. China, in particular, now consumes over half of the world's metals. See World Bank, "Special Focus: How Important Are China and India in Global Commodity Consumption?," July 2015, http://pubdocs.worldbank.org/en/716291444853736301 /CMO-July-2015-Feature-China-India.

4. L. M. Simonian, "Analiz metodologii opredeleniia vybrosov CO_2 na territorii RF premenitel'no k k chernoi metallurgii," *Izvestiia Vysshikh Uchebnykh Zavedenii: Chernaia Metallurgiia* 61, no. 9 (2018): 726, https://fermet.misis.ru/jour /article/view/1407/1151. The original source for these numbers is the *Natsional'nyi doklad o kadastre antropogennykh vybrosov iz istochnikov i absorbtsii poglotitelyami parnikovykh gazov ne reguliruemykh Monreal'skim protokolom za 1990–2016 gg. Chast' 1* (National report on inventory of anthropogenic emissions from sources and removals by sinks of greenhouse gases not controlled by the Montreal Protocol for the 1990–2015. Part 1) (Moscow, 2018), http://cc.voeikovmgo.ru/images /dokumenty/2019/RUS_NIR-2018_v1.pdf; see table 2.4, at p. 23, and table 4.37, at p. 119.

5. In the section that follows, I have benefited from the insightful analysis of Ellie Martus and Stephen Fortescue.

6. "The Changing of the Guard: Shifts in Industrial Commodity Demand," World Bank Commodity Markets Outlook, 2018, http://pubdocs.worldbank.org /en/634041540477574905/CMO-October-2018-Special-Focus.pdf.

7. On the prospects and drawbacks of additive manufacturing, see US National Intelligence Council, *Global Trends 2030: Alternative Worlds* (December 2012), p. 90. For an update, see US National Intelligence Council, *Global Trends: Paradox of Progress* (January 2017). Additive technologies are of increasing interest in Russia, as described in Chapter 7 regarding Rosatom's increasing involvement in this area.

8. Russian customs statistics, reported in Russian Federation, *Rossiiskaia Tamozhennaia Sluzhba* (Federal Customs Service), Moscow, 2020. The annual reports of the Federal Customs Service from 2009 through 2019 are available at https://customs.gov.ru/activity/results/ezhegodnyj-sbornik-tamozhennaya-sluzhba-rossijskoj-federaczii?page=2.

9. KPMG, *Metals and Mining in Russia: Industry Overview and Investment Opportunities,* September 2016, https://home.kpmg/ru/en/home/insights/2016/09/metals-and-mining-in-russia-industry-overview-and-investment-oppor tunities.html.

10. Meenu Gautam et al., "Carbon Footprint of Aluminum Production: Emissions and Mitigation," in Subramanian Senthilkannan Muthu, ed., *Environmental Carbon Footprints: Industrial Case Studies* (Amsterdam: Elsevier, 2018), pp. 197–228.

11. See World Aluminum for 2020 statistics at http://www.world-aluminium.org/statistics/#data. Data for emissions from aluminum are complicated by the fact a portion of aluminum output consists of recycled aluminum.

12. This is true even though the processing of bauxite and alumina are themselves energy-intensive processes, but they are carried out largely outside of Russia. (Russia produces only a minor quantity of the bauxite it uses.)

13. Polina Trifonova, "En+ boretsia s potepleniem," *Vedomosti,* September 24, 2019, accessed via East View (https://dlib.eastview.com/).

14. Martus, "Russian Industry Responses," p. 21.

15. Deripaska website, https://deripaska.com/initiatives/climate-change.

16. Ellie Martus, "Contested Policy-Making in Russia: Industry, Environment, and the 'Best Available Technology' Debate," *Post-Soviet Affairs* 33, no. 4 (2017): 276–297. The article is also available as a chapter in Ellie Martus, *Russian Environmental Politics: State, Industry, and Policy-Making* (Abingdon, UK: Routledge, 2017).

17. For projections to 2030, see World Aluminium (World Aluminium Institute), "Assessment of Aluminum Usage in China's Automobile Industry 2016–2030," http://www.world-aluminium.org/media/filer_public/2019/04/02/execsummary_china_auto_usage_2016-2030_21feb2019.pdf.

18. US Geological Survey, "Mineral Commodity Summaries 2020," p. 21, https://pubs.er.usgs.gov/publication/mcs2020.

19. See the Kremlin website at http://kremlin.ru/events/president/news/59363.

20. For a candid evaluation of the Russian performance to date, see V. A. Kryukov et al., "Rare Earth Industry: How to Take Advantage of Opportunities," *Gornaya promyshlennost',* no. 5 (2020), https://mining-media.ru/ru/article/new tech/16121-redkozemel-naya-promyshlennost-realizovat-imeyushchiesya-voz mozhnosti. However, according to the USGS, Russia has only 10 percent of the world's reserves of rare-earth metals, compared to China's 38 percent.

21. Simonian, "Analiz metodologii." The numbers are somewhat confusing because a portion of the emissions from ferrous metallurgy are reported under "power."

22. Data on Russian metal exports come from Metal Expert (https://metal expert.com/en/index.html). Data are also available from Russian customs statistics and the Russian Statistical Committee and Rosstat.

23. According to MMK's 2019 sustainability report, sales to Russia and the Commonwealth of Independent States (CIS) account for 80 percent of MMK's business.

24. KPMG, *Metals and Mining in Russia.*

25. Fortescue, "Russian Aluminum Industry."

26. I. V. Butorina and M.V. Butorina, "Problemy vnedreniia nailuchshikh dostupnykh tekhnologii v chernoi metallurgii RF," *Chernye metally,* no. 1 (2019): 43–48, https://www.elibrary.ru/item.asp?id=36921232.

27. Kathrin Hille and Henry Foy, "Russia's Next Revolution: How Technology Came to the Mines," *Financial Times,* October 26, 2017, https://www.ft.com /content/addb7e2a-b90d-11e7-9bfb-4a9c83ffa852. (The first half of the article focuses on the coal industry at Kemerovo, the second half on Severstal.)

28. Iuliia Makarova, "Na vstrechu s biznesom," *Izvestiia (Rossiia),* February 12, 2018, accessed via East View.

29. The Global Reporting Initiative originated in 1997 in the wake of the *Exxon Valdez* spill, with the participation of the UN Environment Program. It issues annually updated Standards and Guidelines, which now range well beyond the environment to include issues such as transparency and worker safety. See https://www.globalreporting.org. Interestingly, a search of the website under "Russia" produced no results.

30. Sergei Kashin, "'Zelenye' finansy dlia chernoi metallurgii," *RBK+,* no. 2, July 29, 2019, https://plus.rbc.ru/news/5d3afcf97a8aa91a0c573002.

31. Anatolii Komrakov, "Proizvoditeli stali—ocherednaia tsel' 'zelenyh,'" *Nezavisimaia gazeta,* September 5, 2019, accessed via East View.

32. See the detailed report from McKinsey on various pathways for decarbonizing steel production: Christian Hoffmanh et al., *Decarbonization Challenge for Steel,* June 2020, https://www.mckinsey.com/industries/metals-and-mining/our -insights/decarbonization-challenge-for-steel. The most promising pathway over the long term involves using hydrogen as a reductant.

Conclusion

1. Source: Ministry of Finance, "10-Year Report on Revenues to the Federal Budget," https://minfin.gov.ru/en/statistics/fedbud/.

2. In this connection, see the World Bank study of the "footprint" of the Russian state on the economy: Gabriel di Bella et al., "The Russian State's Size and Its Footprint: Have They Increased?," International Monetary Fund Working Paper 19/53, March 2019, https://www.imf.org/en/Publications/WP/Issues/2019/03/09/The-Russian-States-Size-and-its-Footprint-Have-They-Increased-46662.

3. See the very useful discussion of revenues and expenditures by the Gaidar Institute (Institut ekonomicheskoi politiki imeni E. T. Gaidara), *Turbulentnoe desiatiletie, 2008–2018* (Moscow: Delo, 2020), p. 222. At the Gaidar Institute, Evgeny Goriunov and his colleagues have pioneered the study of Russia's coming budget gap, notably in Evgeny Goriunov et al., "Russia's Fiscal Gap," National Bureau of Economic Research Working Paper 19608, November 2013, https://www.nber.org/papers/w19608. The outlook for Russia's likely budget gap is expanded in a report by Janis Kluge, "Mounting Pressure on Russia's Government Budget," SWP Research Paper No. 2 (Berlin: Stiftung Wissenschaft und Politik, February 2019), https://www.swp-berlin.org/10.18449/2019RP02/.

4. Gaidar Institute, *Turbulentnoe desiatiletie, 2008–2018*.

5. M. Saunois, "The Global Methane Budget," *Earth System Science Data* 12, no. 3 (2020), https://essd.copernicus.org/articles/12/1561/2020/.

6. The "southern" region is defined here as the "Southern Federal Region" (*okrug*). See *Rossiiskii Statisticheskii Ezhegodnik* (Moscow: Rosstat, 2019), p. 93.

7. In 2018, 110 million of Russia's 147 million inhabitants lived in cities, of which 73 million lived in cities with populations over 100,000. See *Rossiiskii Statisticheskii Ezhegodnik*, p. 103. It is true that some large cities, such as Omsk and Yakutsk, are located in the Arctic, but most are not.

8. Generation Z is commonly defined as including those born between the mid- to late 1990s and the early 2010s. For a sampling of what the future might bring from today's Generation Z, see https://trends.rbc.ru/trends/green/5ef479d89a7947afd8d5a4bd and https://trends.rbc.ru/trends/green/5f21b22b9a794773f0368ab1.

9. On the state of the Russian economy on the eve of the COVID-19 pandemic, see the annual reports of the Russian office of the International Monetary Fund. The 2019 report can be accessed at https://www.imf.org/en/Publications/CR/Issues/2019/08/01/Russian-Federation-2019-Article-IV-Consultation-Press-Release-Staff-Report-48549. The report commends the Russian government for its sound macroeconomic policy, consisting of "moderately tight monetary policy with a broadly neutral fiscal stance."

10. See Anders Aslund, *Russia's Crony Capitalism: The Path from Market Economy to Kleptocracy* (New Haven, CT: Yale University Press, 2019). Despite the book's title and its overall focus, chapter 3 contains a description of Russia's overall conservative monetary and fiscal policies under Putin. The main architect of these policies was Aleksei Kudrin, whom Putin kept in place as his finance minister for

a decade. Key members of Kudrin's team remain in charge of policy today. Russia's government shift in early 2020 brought to the post of prime minister one of the architects of Russia's fiscal modernization, Mikhail Mishustin. Putin's financial team also includes German Gref (now the head of Sberbank) and El'vira Nabiullina (head of the Russian State Bank), both of whom, together with Kudrin, came with Putin from Saint Petersburg in the early 2000s.

11. Russian economists like Iakov Mirkin agonize over Russia's slow GDP growth rates. Russia's GDP grew by only 15 percent between 2010 and 2019, Mirkin complains, versus 88 percent for China and even 22 percent for the United States. "Rapid growth is a matter of technology." Iakov Mirkin, "Pochemu Rossiiskaia ekonomika rastet tak medlenno," *Vedomosti,* July 16, 2020, https://www.vedomosti.ru/opinion/articles/2020/07/16/834679-rossiiskaya-ekonomika.

12. A powerful book that argues this point is Clifford G. Gaddy and Barry W. Ickes, *Bear Traps on Russia's Road to Modernization* (Abingdon, UK: Routledge, 2013).

13. Kathrin Hille and Henry Foy, "Russia's Next Revolution: How Technology Came to the Mines," *Financial Times,* October 26, 2017, https://www.ft.com/content/addb7e2a-b90d-11e7-9bfb-4a9c83ffa852. To Severstal's credit, the plant at Cherepovets and the surrounding city are being extensively modernized, and they are already much changed from their Soviet past.

14. In 2014 the government created a Fund for the Development of Monotowns, which pledged $452 million to improve their economies and attract investors, but this program has been hampered by the economic slowdown of the last decade and the advent of COVID-19.

15. "A massive survey by the Centre for Strategic Research (CSR), a think tank headed by Kudrin, found that the average age of the equipment in Russian oil production is 19 years, in metallurgical plants 17 years and in mining of resources other than oil 16 years." https://www.csr.ru/upload/iblock/f37/f372bf13f0289cdba a86071ac4451408.pdf.

16. Eastern Economic Forum, September 2019, http://kremlin.ru/events/pres ident/news/61451.

17. "Russia: Exports, Percent of GDP," TheGlobalEconomy.com, https://www.theglobaleconomy.com/Russia/Exports/. It is better to use turnover as a share of GDP. Turnover in 2019 was $672 billion (Federal Customs Service, http://customs.gov.ru/statistic/vneshn-torg/vneshn-torg-countries). GDP (nominal) was $1.7 trillion.

18. Harley Balzer, "Russia and China in the Global Economy," *Demokratizatsiya: The Journal of Post-Soviet Democratization* 16, no. 1 (2008): 37–47, https://demokratizatsiya.pub/archives/16_1_R91771QK24UJ753L.pdf.

19. Balzer, "Russia and China in the Global Economy."

20. World Bank, "Foreign Direct Investment, Net Inflows (% of GDP)—Russian Federation," https://data.worldbank.org/indicator/BX.KLT.DINV.WD.GD.ZS?end =2018&locations=RU&start=1992&view=chart. Data on FDI for Russia must be interpreted with caution—Russian investors habitually move capital out of Russia to countries that give them favorable treatment, then repatriate capital back into Russia as "direct investment" when conditions improve. For a discussion, see BOFIT (Bank of Finland), *BOFIT Weekly* no. 22 (2019), https://www.bofit.fi/en/monitoring /weekly/2019/vw201922_2/. Nevertheless, the overall pattern is clear.

21. Filip Novokmet, Thomas Piketty, and Gabriel Zucman, "From Soviets to Oligarchs: Inequality and Property in Russia, 1905–2016," *Journal of Economic Inequality* 16, no. 2 (August 2017): 189–223, https://link.springer.com/article/10 .1007/s10888-018-9383-0. An earlier version of this article is available as a research paper of the National Bureau of Economic Research (NBER) at https://www.nber .org/papers/w23712.pdf.

22. Timothy J. Colton, *Russia: What Everyone Needs to Know* (Oxford: Oxford University Press, 2016), pp. 7, 12.

23. A fine recent overview of Russia's military reforms and their impact on the development of new weapons is Bettina Renz, *Russia's Military Revival* (Cambridge: Polity Press, 2018). Professor Renz, who teaches at the University of Nottingham, has her roots at the University of Birmingham's Centre for Russian and East European Studies, which for many decades has been the West's leading center of excellence in the study of Soviet and Russian technology.

24. Despite this, Russia's new military prototypes are impressive. For example, the new Tsirkon hypersonic missile, tested in October 2020, can travel at eight times the speed of sound and hit naval targets up to 4.5 kilometers away. The first successful test was reported to President Putin on his birthday. See the story on TVZvezda, "Fregat 'Admiral Gorshkov'" vypolnil strel'bu raketoi 'Tsirkon' iz Belogo Moria," *TVZvezda*, October 7, 2020, https://tvzvezda.ru/news/forces /content/2020107102- 1EJJs.html?utm_source=tvzvezda&utm_medium=longpage &utm_campaign=longpage&utm_term=v1. But significantly, these are still considered prototypes.

25. This is also the argument of Anders Aslund in the concluding chapter of his excellent book *Russia's Crony Capitalism: The Path from Market Economy to Kleptocracy* (New Haven, CT: Yale University Press, 2019), pp. 226ff. Aslund concludes, "Russia is a classical declining power."

26. I have borrowed this nice phrase from Professor Timothy Colton (private communication with the author), with thanks.

ACKNOWLEDGMENTS

This book is testimony to the power of friendship and collaboration. Much of the research and writing was done as the COVID-19 pandemic spread across the world, and many of us, including the author, were confined at home. Yet we discovered that we could reach out across countries and continents and continue friendly and productive contact. My heartfelt thanks go out to all of you who contributed to this project.

First and foremost, I thank my longtime friend and colleague—and former student—Philip Vorobyov, to whom this book is dedicated. Phil read and reread successive drafts of each chapter and, over the course of many evening conversations from his home in London, gave generously of his advice and support. Many thanks to you, Phil!

My Georgetown colleagues Harley Balzer and Marjorie Mandelstam Balzer were powerful inspirations throughout. They provided ideas and suggested sources, read and commented on drafts, and shared with me their vast network of contacts and friendships with leading Russian scholars.

No less valuable was the help of Alla Baranovsky of Harvard University, who acted as editor, critic, and overall guide and advisor, as she read draft after draft, checked sources, and suggested many improvements in style and content along the way. Many thanks to her, and to my friend Timothy Colton for his inspired move in bringing us together.

Bruce Parrott of Johns Hopkins SAIS, a close friend since Columbia days in the 1960s, was a constant source of encouragement and friendly advice, as he too read and commented on successive drafts. Similarly, Simon Blakey, who was in many ways the de facto coauthor of my previous book, *The Bridge,* was an indispensable source of support in the new project, especially on natural gas and LNG.

A small army of friends and colleagues read chapters and advised on individual fields in which they were much better qualified than I. On oil: Keith King, Maksim Nechaev, Adnan Vatansever, Chris Weafer, and John Webb. On the Arctic: Valeriy Kryukov, Arild Moe, and Robert Orttung. On natural gas and LNG: Fu Jingyun, Anna Galtsova, James Henderson, Laurent Ruseckas, Jesse Scott, Jonathan Stern, Michael Stoppard, and Vitaly Yermakov. On nuclear power: Paul Josephson and Scott Stephenson. On policymaking: Vadim Grishin, Emily Holland, Peter Kaznacheev, Robert Otto, and Peter Reddaway. And special thanks to Stephen Wegren, who coached me on the extraordinary revival of Russian agriculture, on which he is the West's leading expert.

Russian friends and colleagues were also a valuable source of advice. Many of them are part of the increasingly powerful wave of awareness and activism on climate change in Russia. In addition to those already mentioned, they include Oleg Anisimov, Irina Dezhina, and Tatiana Mitrova.

I am grateful to my former teammates at IHS Markit for many years of close partnership and friendship. Daniel Yergin, as always, is the group's indispensable guide and mentor. The Caspian and Russian Energy team continues to prosper, under the able leadership of Matt Sagers, Irina Zamarina in Moscow, and my former student Dena Sholk in Kazakhstan. I wish them all Godspeed in their new home with S&P Global.

Georgetown University, though physically closed because of COVID-19, remains a vital force for education and scholarship, and I thank my students and fellow faculty members for their support and inspiration, mostly delivered via Zoom. I am grateful to our Government Department chairmen Charles King and Anthony Arend for

Acknowledgments

their encouragement as I developed a new field of teaching on climate change, and to my friend and colleague Angela Stent, an outstanding teacher and author on Russia's foreign relations, who heads our CERES program on Eastern Europe and Russia. Finally, I owe a special debt to Lauinger Library and to the unsung heroes of the Interlibrary Loan program, Dana Aronowitz and Amanda Rudd.

This is my third book with Harvard University Press, which has been a happy and creative home for over a decade. I am especially grateful to my talented editor, Janice Audet, to her past colleagues Mike Aronson and Jeff Dean, and to the team that has provided such excellent support in publicizing and promoting *Wheel of Fortune, The Bridge,* and now *Klimat*—Eleanor Andrew, Emeralde Jensen-Roberts, Megan Posco, and Karen Pelaez. Jerome Cookson created the maps and graphics, as he did for *The Bridge,* with unequaled skill. John Donohue of Westchester Publishing Services was again the ever-patient and sharp-eyed production editor, and Wendy Nelson was the invaluable copyeditor. Many thanks to all of them.

And lastly, as always, this book owes its existence to the constant warm support of my family, Nil, Peri, Farah, and Kenan, to whom I am grateful every day.

INDEX

Note: Page numbers in *italics* indicate figures and tables.

94, 253n44, 253n47, 253n49; nuclear power in, 135, 137–138, 140, 143–144, 146–147, 150–152, 272n65, 273n69; renewables industry in, 114, 121–122, 128, 263n35, 264n43; role of in metals trade in, 199–200, 202–203, 286n3, 287n20

Chubais, Anatoly, 43; as advocate of Russian renewables, 113, 117–123, 126–127, 129–130, 260nn3–6, 261n8, 264n45, 265n57; changing views on nuclear power, 133, 267n7; as "father of Russian privatization," 112; as former head of Rusnano, 23, 113, 260n4; as former head of UES, 112, 118; and Putin, 23–24, 113–114, 232n18, 260n6, 265n57

Chukchi Sea, 150

Climate change in Russia: activists and, 23, 43–46, 84; benefits of for Russia, 5–6, 14, 208, 224, 284n28; as century's defining issue, 1, 207; as collective action problem, 2; consequences of for Russia, 1–3, 6, 14–16, 207–209, 224, 284n28; COVID-19 and, 1, 9–11, 228n6; effects of (direct vs. indirect, internal vs. external), 6, 14–16, 208, 214; as elite issue, 3, 17–24, 43, 216; as evolving global issue, 2, 4, 31; four categories of players in, 17–24; global diplomacy of, 219, 221; impact on Generation Z, 215, 289n8; interaction with coming political succession, 207, 217; interaction with economic issues, 2–3, 6, 207–208, 214; interaction with other political issues, 2–3, 14, 45, 207–208, 209, 214–215; public opinion on, 14, 17, 20, 31, 35, 43–46, 93, 181, 187, 214

(*see also* Public opinion in Russia on climate change); Russia as cause and victim of, 5, 209, 227n1; vs. United States, 7–9, 27, 45, 50, 209, 227nn3–4, 227–228n5 (*see also* United States); vulnerabilities to, 5, 46, 189, 202

Climate Change Adaptation Plan, 33, 236n46

Climate "conservatives," 36, 39, 60, 216; positions of, 17, 19–21; Putin as, 19–20, 24, 32, 34, 38, 43, 47, 96

Climate diplomacy, 18, 21, 155, 214, 219, 221

Climate Doctrine of 2009, 28–30

Climate scientists, 11; Charles Keeling (Keeling's Curve), 25, 232n21; early role in USSR, 24–26, 232–233n22; Mikhail Budyko, 25, 232–233n22, 233n24; in Russia, 14, 17–18, 229n2; in United States, 24–25

CNNC New Energy, 150, 273n69

Coal. *See* Russian coal

Coastal erosion, 186, 194, 282n12

Collectivized agriculture, 153–154, 163, 274n1. *See also* Kolkhoz; Sovkhoz

Colombia, 99, 109

Colton, Timothy, 222, 291n26

CO_2 emissions, 8, 11, 19, 29, 50, 147, 206, 214, 230n6, 286n4, 287n11, 288n21; from agriculture, 155–156; from coal, 96–98, 108, 110; cuts in, 2, 10, 13, 21, 26, 28, 36, 40–41, 201–202, 221; early observations of rise of, 24–25; growth in, 1–2, 13–15, 96; "net zero" goal, 51, 101; reporting of, 30, 36, 204–205; in Russia, 4–5, 27, 33–34, 40, 42–43, 77, 114, 126, 129–130, 197–198, 200, 230n6, 233n26, 235n38, 237n61

Permian basin, 64

Pevek, 150, 185

Philippines, 108

Pollution, vs. climate change, 10, 34–35, 44, 46, 50, 79, 203, 214, 236n45; from coal, 79–80, 97, 101, 103, 111; from other sources, 90, 144, 181, 202–203

Potanin, Vladimir, 46, 181, 202. *See also* Nornikel'

Powering Past Coal Alliance, 81

Power of Siberia (PoS) gas pipelines, 80, 85–88, 250n20

Prodimex, 167

Public opinion in Russia on climate change, 14, 20, 31, 35, 43–46, 93, 181, 187, 214

Putin, Vladimir, 3, 21, 45–46, 59–60, 93, 234nn30–31; on agriculture, 157–158, 168–170, 173–176, 179–180, 278n49, 279n50, 279n54; and Anatoly Chubais, 23–24, 113–114, 232n18, 260n3, 260n6, 265n57 (*see also* Chubais, Anatoly); and the Arctic, 181, 190–192, 284n25, 285nn37–38; backing for LNG, 73–74; as climate conservative, 19–20, 24, 32, 34, 38, 43, 47, 96; and coal, 96–97, 107, 110; dominance of "hydrocarbon model" under, 4–5, 24, 36–37, 48, 75, 213, 215, 218; efforts of to promote new blood at top of government, 21–23; long-term legacy of, 207, 215, 220, 223, 289–290n10; on metals, 202, 204; on methane emissions and Venus, 37–38; on nuclear energy, 37, 131, 140, 142–143, 147, 149–151, 267n4, 270n42, 273n71; and the oil industry, 47–48, 53, 59–60, 65–66, 244nn39–40; positive achievements

under, 215–216, 289–290n10; rejection of peak oil demand narrative, 34; relationship with RSPP, 39–41; on renewables, 36–37, 114, 118, 127, 260n3; and Russia's "Pivot to the East" strategy, 85; sources of climate-change views of, 38; "two Putins" on climate change, 34–38; views on climate change, 23–24, 27–28, 31, 33–38, 43, 233n28; views on Rosneft and Gazprom, 47, 53, 235n41, 239n2, 248n8

Qatar, 78, 86, 94, 249n13, 250n25, 251n26

Reclamation of land, 155, 159, 162, 164, 172, 176–178, 280n64

Red Wind, 124

Renewable energy. *See* Russian renewable energy

Renova, 114, 120–121, 263n32, 263n35. *See also* Vekselberg, Victor

Rents from oil and gas, 68, 76, 207, 212, 216, 244n39

Reshetnikov, Maksim, 21, 230n6

RIF, 170–171, 278n42

Romania, 118, 146, 267n8

Rosatom: burnishes green credentials with ESG sustainability statements, 147; development of new technologies by, 19, 148, 151–152, 286n7; and floating nuclear power plants, 149–150, 273n68; history of, 268n20, 270n33, 271n49, 271n51; international business of, 124, 140, 143–147; military activities of, 141–142, 273n71; Northern Sea Route and, 124, 147–148, 192–193, 195; as parent company of Rosatomflot, 149–150, 192–193; as promoter of green

Rosatom (*continued*)
   virtues of nuclear power, 19; and
   renewable energy, 113, 119, 123–125,
   127–129, 152, 265n49, 265n51,
   272n62; restructuring of under
   Sergey Kirienko, 131–132, 139–142,
   272n59. *See also* Kirienko, Sergey
Rosatomflot, 149–150, 192–193
Rosgidrometsluzhba (Rosgidromet),
   18, 29–31, *156*, 234n34
Rosneft, 19, 47–48, 53, 58, 60–61, 65–66,
   87, 193, 235n41, 239n2, 246n53,
   248n8, 278n42. *See also* Sechin, Igor
Rostov Oblast', 119, 124–127, 170–171,
   265n56, 276n29
Rusagro, 167
Rusal, 196–198, 200–202, 235n38
Rusatom-Additive Technologies
   (RusAT), 148
Rusnano, 23, 113–114, 119–121, 126–127,
   130, 232n17, 260n4, 260n6, 263n34,
   264n37, 266n61, 267n7
Russia: achievements under Putin in,
   215–216, 289–290n10; demographic
   trends in, 7, 69, 113, 117, 122, 161,
   184–187, 194, 211, 214, 218, 221,
   281–282n10; energy transition in,
   7, 9, 11–14, 21, 46, 48–52, 77, 89–91,
   132, 147, 150, 208–209, 213, 228n9,
   229–230n4, 234n32, 241nn11–12;
   future position of as great power,
   2–3, 221–223; generational change
   in, 21–23, 108, 151, 168, 215, 223–224;
   "hydrocarbon model" in, 4–5, 188,
   194, 207, 210, 218, 224; nonenergy
   exports from, *15*, 16; as "resource
   appendage" (*resursnyi pridatok*),
   136–138; strengths and weaknesses
   in science and technology (*see also*
   Technology in Russia); "thick" vs.

"thin" integration in world economy,
   4, 10–11, 136, 208, 214, 219–221; vs.
   United States, 31–32, 128, 220–221,
   247n57, 249n9
Russian Academy of National Economy
   and Public Administration
   (RANEPA), 166, 170, 175, 260n50
Russian agriculture, 14; agro-holdings
   in, 154, 157, 165–176, 178, 278n42,
   278n45; boom in Russia, 4, 153–155,
   162–165; compared to United States,
   8, 153, 179, 277n40; countersanctions
   in, 158, 168, 274n4, 276n18; crop
   failures in, 2, 31, 35; declining imports
   in, 154, 158, 162–163; dispossession
   by "raiders" in, 164; drought in, 5,
   8, 14, 35, 153, 156, 158–160, 162, 171,
   175, 177, 179, 213–214, 275n13; exports
   from, 4, 153–155, 157, 159, 162–163,
   165, 168, 170–171, 177–179, 213;
   floods in, 7–8, 14, 153, 159, 172,
   175, 186–187, 227n2, 282n13; grain
   production in, 157–159, 161–163,
   166, 170–171, 173–174, 178, 213, 274n5,
   275n17, 276nn29–30, 277n40,
   278n42; heat waves, 2, 14, 20, 31,
   159, 214; impact of climate change
   on (two phases of), 5, 155, 158–159;
   intensification policy (*intensifi-
   katskiia*), 162; Non–Black Earth
   zone (*Nechernozem'e*) in, 154, 159,
   161; privatization of, 154, 165–166,
   173; reclamation in, 155, 159, 162,
   164, 172, 176–178, 280n64; role of
   state in, 154–155, 158, 161–164, 166,
   172, 173–176, 178–179, 215; as source
   of greenhouse gas emissions (CO_2
   and methane), 155–156; under
   USSR, 153–154, 157, 161–163, 171,
   174–175, 176, 179, 274n1, 276–277n30

Russian aluminum: Chinese market for, 202; contribution of to global CO_2 emissions, 200–201; dependence on imported alumina in, 200; dominant role of Rusal in, 200; proactive role in climate change policy, 201–202. *See also* Deripaska, Oleg; Rusal

Russian Arctic, 14, 39; coastline of, 5–7, 54, 65–67, 124, 149–150, 183–186, 188, 190–191, 194–195, 212, 231n12; as example of dominant "hydrocarbon model," 188, 191, 194; government's heightened focus on (from 2019), 33–34, 183–184, 188, 191, 195, 236n47; government's "Two-Track" policy in, 188–189; growing role of Rosatom in, 192–193, 195, 286n7; growth of Yakutsk in, 186–187, 194, 218, 282n14, 289n7; Indigenous Peoples, 186, 187, 218; Krasnoyarsk Kray in, 22, 185, 189, 231n12, 242–243n24, 247n58; neglect and stagnation of inland of, 183–188; Northern Sea Route in, 5, 14, 34, 66, 74, 81, 148, 184, 188–193, 195, 212, 283n22, 283n24, 284n25, 284n27, 284n34; nuclear icebreakers in, 74, 148, 189, 192–193, 195; oil and gas in, 184–185, 187–188, 190–191; permafrost in, 5, 7, 14, 20, 34, 37, 46, 76, 160, 177, 181–187, 194, 212–214, 217, 229n2, 4, 281n3, 282n14, 285n40; in Russian export strategy ("pivot to the East"), 76, 85, 183–184, 190–191, 194–195; Soviet treasure-house, 185; two narratives, 183–184; Yamal Peninsula in, 74–76, 80, 86, 88, 185, 189, 195, 215, 235n39, 249n15

Russian coal, 12–14; "clean coal," 84, 106; coking coal, 105, 258n30; compared with United States, 98, 101; demand in China for, 12–13, 79, 100–103, 106, 109, 111, 229n14, 260n52; demand in the West for, 98–100, 110; demand in the world for, 12–13, 208–209; dependence of on railroad transportation, 95, 102, 104–105, 107, 110; domestic demand for, 97–98, 199; emissions and, 96–98, 108, 110; natural gas and, 98, 106; renewables and, 100, 103, 111, 210; role of in Russian export revenues, 96–97; seaborne coal, 109; in USSR and after, 4, 104–106, 215; steam coal, 99, 105, 258n30; two phases to Russian coal strategy, 103

Russian Council of Industrialists and Producers, 108

Russian economy: default and ruble devaluation, 131, 266n1; financial crisis of 2008, 143; hydrocarbon model in, 4–5, 188, 194, 207, 210, 218, 224; integration into global economy, 4, 10–11, 136, 208, 214, 219–221; modernization of, 4, 28; oligarchs and, 88, 112, 120, 127, 140, 165, 181, 196–197; privatization of, 53, 73, 104, 112, 120, 154, 165–166, 173; "resource appendage," 136–138

Russian Hydrometeorological Service. *See* Rosgidrometsluzhba (Rosgidromet)

Russian metals industry, 14; and China, 199–200, 202–203, 286n3, 287n20; contrast with Russian coal industry, 198–199, 201–203, 205–206; "green credentials" of, 197, 199, 201, 204; post-Soviet modernization, 4,

Russian metals industry (*continued*) 196–197; potential impact of CBAM on, 198, 201, 205–206; role of in CO_2 emissions, 197–198, 200–203

Russian military, 190–192, 195, 197, 217, 222, 285n37, 291nn23–24; nuclear and, 141–142, 149, 273n71; Soviet legacy in, 138–139, 192, 221–222

Russian natural gas, 14; Chinese demand for, 74, 76, 80–81, 83–89, 94, 252n44, 253n47; and CO_2 emissions, 77; vs. coal, 84, 86–87; domestic consumption of, 76–77, 116, 250n20; in electric power sector, 76–77, 81, 84; "embodied gas," 87–89; future markets for, 77–81, 83–87, 208–210; liquefied natural gas, 4–5, 15, 22, 34, 73–74, 76, 78–81, 83–84, 86–88, 94, 97, 150, 169, 184, 189, 191, 193, 195, 209, 212, 215, 242n18; pipelines, 15, 73–74, 76, 78, 80, 83–88, 91, 106, 215, 242n15, 248n2, 249n15, 250n20, 251n30, 251n44, 255n63; "pivot to the East" policy and, 76, 85; vs. renewables, 80, 82; reserves of, 4, 75–77; shale gas from United States and, 32, 78, 83, 93, 115, 240–241n10; under USSR, 4, 73, 75–77, 210, 215

Russian nuclear power, 14; advantages of, 133–134, 147; challenges of, 135, 145, 147–149; vs. China, 135, 137–138, 140, 143–144, 146–147, 150–152, 271n58, 272n65, 273n69; contrasted with United States and France, 134, 137–139, 144, 152, 290n11; "cookie cutter" strategy in, 144; domestic nuclear market, 142–143, 147, 151; floating nuclear power plants, 146, 149–152, 272–273n66; "full service" package in, 144; history of in USSR, 138–140, 143, 149, 189, 192, 269n28, 283n22; international position of, 132, 140–145, 147, 151, 209; military origins of, 141–142; renaissance of, 4, 132, 140–142; role of Sergey Kirienko in (*see* Kirienko, Sergey); small modular reactors in, 135, 152, 268n17, 273n74; VVER as main technology in, 139, 144–146. *See also* Rosatom

Russian oil, 14; Arctic clusters and, 66; "brownfield" vs. "greenfield" vs. "bluefield" oil in Russia, 54–56; compared with United States, 32, 61, 63–65, 245n47; dependence of Russian state budget on exports of, 47–48, 52, 60, 67–68, 198, 211; hard-to-develop reserves (*trudnoizvlekaemye zapasy*, or TRIZ) in, 57; impact of Western sanctions on, 56, 59, 247n57; inadequate domestic technology in, 54, 56, 58–59; lack of funding for exploration for, 54, 56; peak demand for, 6, 14, 22, 32, 34, 51–52, 56, 62, 75, 230–231n11, 240n9; policy, 52, 56–57; post-Soviet modernization of, 4, 53, 215; tax relief and subsidies in, 59–60, 62, 106; "tight oil" (shale oil) in, 12, 32, 54, 61, 63–66, 71, 245n47, 246nn50–51; in USSR after the breakup, 4, 52–55, 58, 210, 242nn22–23; West Siberia as declining workhorse of, 54

Russian railroads (Rossiiskie Zheleznye Dorogi, or RZhD), 69, 152, 185, 188; critical role of in coal exports, 95, 102, 104–105, 107, 110; "Eastern Polygon" program, 107, 259n41; as state-owned company, 105, 107

Russian renewable energy, 9, 11, 14, 22, 46; Anatoly Chubais and, 23, 112–114,

Turkmenistan, 83, 249n13
Tuzov, Vladimir, 108–109, 259nn45–46.
    See also SUEK
Tyumen Oil Company (TNK), 120

Ukraine, 32, 76, 138, 144, 167, 220,
    249n15
Ul'ianovsk Oblast, 126
United Machine Building Factories
    (OMZ), 140
United Nations Climate Change Con-
    ference, 18, 110
United Nations Food and Agriculture
    Organization (FAO), 155
United Nations Human Development
    Index, 12
United Power Systems (UES), 112,
    118, 260nn3–4. *See also* Chubais,
    Anatoly
United Russia Party, 168
United States, 36; agriculture in, 8, 153,
    179, 277n40; climate action in, 27,
    45, 50, 209, 227–228n5; climate
    scientists in, 24; coal in, 98, 101;
    natural gas in, 32, 78, 83, 93, 115,
    240–241n10; nuclear power and
    technology in, 134, 137–139, 144,
    152, 290n11; oil industry in, 32, 61,
    63–65, 245n47; relationship with
    China, 219; relationship with Russia,
    31–32, 128, 220–221; renewables
    in, 80, 122, 128; vulnerabilities to
    climate change in compared to
    Russia, 7–9
Ural Mountains, 113, 184
USSR (Union of Soviet Socialist
    Republics): climate science in,
    24–26; coal industry in, 4, 104–106,
    215; collectivization and agriculture
    under, 153–154, 157, 161–163, 171,

174–175, 176, 179, 276–277n30;
    economy in, 27, 185, 196–197, 207,
    221–223, 290n13; industrialization
    under, 217–218; "monocities" in,
    217–218; natural gas in, 4, 73, 75–77,
    210, 215; nuclear industry and
    technology in, 138–140, 143, 149,
    189, 192, 283n22; oil industry in, 4,
    52–55, 58, 210
Uzun, Vasily, 166, 169, 175

Valdai Conference, 37
VEB.RF, 23, 106, 114, 230n8, 232n17,
    260n6
Vekselberg, Victor, 120. *See also*
    Renova
Vestas, 126, 129
Vietnam, 108
Vogtle 3 and 4, 134
Volga-Urals basin, 54, 104
Volgo-Don Invest, 168
Vorkuta, 185, 217
Voronezh, 168, 276n29
VTB, 106

Wegren, Stephen, 157, 163, 274n4,
    275n17, 277n32, 280n69
Wheel of Fortune (Gustafson), 54,
    242n22, 248n8, 269n25
Wind power, 265n49, 51, 52; declining
    costs worldwide, 114–116; growing
    role of Rosatom through wind sub-
    sidiary NovaWind, 124; offshore
    wind, 127–129; reasons for Russian
    lag in, 123. *See also* Russian renew-
    able energy
World Bank, 30, 104, 199, 219, 235n41
World Trade Organization, 27, 42
World Wildlife Fund (Russia) (WWF),
    42, 227n2